Scientific Failure

Scientific Failure

Edited by Tamara Horowitz
and Allen I. Janis

Rowman & Littlefield Publishers, Inc.

ROWMAN & LITTLEFIELD PUBLISHERS, INC.

Published in the United States of America
by Rowman & Littlefield Publishers, Inc.
4720 Boston Way, Lanham, Maryland 20706

3 Henrietta Street, London WC2E 8LU, England

Copyright © 1994 by Center for the Philosophy of Science

British Cataloging in Publication Information Available

Library of Congress Cataloging-in-Publication Data

Scientific failure / edited by Tamara Horowitz and
Allen I. Janis.
p.   cm.
Includes bibliographical references.
1. Science—Philosophy.   2. Research—Methodology.
3. Science—Social aspects.   I. Horowitz, Tamara.
II. Janis, Allen I. (Allen Ira).
Q175.S4234   1993   502.8—dc20   93–5110 CIP

ISBN 0–8476–7806–7 (cloth : alk. paper)

Printed in the United States of America

Contents

PREFACE

This volume was inspired by a workshop on "Scientific Failure," which was organized by Gerald Massey and us, sponsored by the Center for Philosophy of Science at the University of Pittsburgh, and held at the University on 23 and 24 of April 1988. The workshop brought together philosophers, historians, and scientists from a number of disciplines to examine the role of scientific failure in the development of science.

Some of the papers in this volume were among those presented at the workshop, others were written specifically for this volume. We are grateful to all of the participants at the workshop as well as the other contributors to this volume for joining in the effort to focus attention on scientific failures instead of the more common concentration on scientific successes. We hope that this volume will stimulate a wider community of philosophers, historians, and scientists to examine the issues raised herein.

We are grateful to many people for their help with both the workshop and this volume. In particular, we wish to thank Gerald Massey, who at the time of the workshop was Associate Director of the Center and shortly thereafter became Director, not only for his role in organizing the workshop but also for his assistance in the early stages of assembling this volume. We also thank Nicholas Rescher, who was Director of the Center at the time of the workshop and is now its Vice Chair, for his encouragement and support. Linda Butera helped with arrangements for the workshop, and Mary Connor helped with some of the organizational tasks associated with this volume. Neither the workshop nor this volume would have been possible without financial support from the R. K. Mellon Foundation.

Tamara Horowitz & Allen I. Janis
Pittsburgh, Pennsylvania
August 1992

INTRODUCTION

Both philosophers and scientists, when discussing scientific progress, almost always seem to concentrate on the successes of science. A notable exception was the philosophically minded physicist Niels Bohr, who liked to talk about the two types of truth: An ordinary truth was one whose negation was certainly false; a deep truth, on the other hand, was one whose negation was also a deep truth. What Bohr had in mind was the struggles he and the other founders of quantum theory had with its interpretation. What, for example, did it mean to speak of wave-particle duality? That a quantum of light was a wave was a deep truth; that it was a particle was also one. Bohr's deep truths can indeed be viewed as failures of science, at least at the time that the deep truths are the best that one can do. Yet Bohr saw the realm of deep truths as an important, perhaps even necessary, intermediate step toward reaching true understanding. It was only through wrestling with those apparent inconsistencies that one achieved a fuller comprehension of the true nature of things.

There are indeed many instances where an examination of the failures of science sheds light on its development or, in the case of contemporary failures, provides insights that help one see one's way out of the morass. The papers in this volume are offered in the hope that they will stimulate both philosophers and scientists to think more about the failures of science, and thereby better understand its successes.

As an overview of the volume, we present here brief summaries of each of the papers it contains.

SECTION I—METHODOLOGY: THE ROLE OF FAILURE IN MODERN SCIENTIFIC DEVELOPMENT

"The Value of Scientific Failure"
Allen I. Janis

This first paper of the volume illustrates the book's theme with a discussion of the scientific value that can be derived from three distinct types of scientific failure. Each of these types is illustrated with a particular example, showing how science benefited from that failure.

The first type is that which is most often tied to scientific progress. A theory that has previously been considered to be adequate is now found to be unable to account for new observations. In the case of a well-established theory, a common first response to such observations is to question them. But if they indeed hold up, attempts are made to alter the theory or find a competing theory that will better account for the totality of observations.

The history of science provides many examples of this sort of failure and the subsequent progress that ensues. The particular example used by Allen I. Janis is from the fairly recent history of superconductivity when superconductivity was found to take place at temperatures much higher than should have been possible under the then accepted theory.

A familiar piece of folklore is that one learns from one's mistakes. Although most people would subscribe to this view, it is nevertheless also a common view that it is even better not to have made the mistakes in the first place. Janis's second type of failure is meant to illustrate that science would, in some cases, have progressed more slowly had the failure not occurred. This type of failed theory is one that was never accepted as probably true (or as true as theories are ever viewed as being). Yet such theories can nevertheless benefit science if they lead, more or less as by-products, to significant advances in related fields. The example Janis gives here is the steadystate theory of cosmology, a theory that at best was never considered to be more than one of a number of possible alternatives, and is now considered by the vast majority of cosmologists to be wrong. It led, however, to a better understanding of the formation of elements, an understanding that undoubtedly would have come more slowly had the steadystate theory never been proposed.

The third type of failure is of a different sort—a failure of practice rather than of theory. This is the failure to take account of all relevant data when comparing a theory with experiment. That such failure can sometimes be beneficial is illustrated by Janis's example of Einstein's early treatment of the specific heat of solids. This theory is viewed by many physicists as the beginning of the modern theory of condensed-matter physics, and it led to many fruitful developments. Yet had Einstein been aware of then existing data that contradicted his theory, he may well have refrained from publishing his ideas. Einstein's failure to be complete may again have hastened the progress of science.

"The Frustrated State of Nonequilibrium Statistical Physics"
James V. Maher

Although nonequilibrium statistical mechanics has been investigated for about a century, James V. Maher considers the theory, in its present state of development, to be an example of scientific failure. Most physicists would agree that it has indeed had few successes, yet much effort still goes into attempts to study and understand nonequilibrium phenomena; Maher himself does experimental work in this area. One reason that this battle for progress continues is the importance of the problems encompassed. Maher lists a number of them. A thorough understanding of turbulent systems, for example, could make it possible to have reliable weather forecasts; and since living organisms are certainly far from a state of equilibrium, the question of whether biology can be reduced to physics cannot even be approached without knowing whether nonequilibrium statistical physics can be understood. Yet these and other basic problems remain unsolved, leading to the view that the theory is, at least so far, a failure.

Maher's analysis of the responses of the physics community to this failure distinguishes three phases. The first is denial: There are no problems of principle, just difficulties in carrying out the necessary calculations. But the development of the theory of chaos has made clear that there are indeed problems of principle in at least some cases. Then comes name calling: Blame it on the failure of mathematicians, who have not provided physicists with a good theory of nonlinear differential equations. And finally, dig in and see what can be done: With the advent of more powerful computers and newer mathematical techniques, there is now a flurry of renewed attacks on these difficult nonequilibrium phenomena. Maher sees cause for hope that the physics is on the right track, so that progress in mathematics and computation will indeed lead to progress in understanding.

Returning again to the problem of principle raised by chaos, we must realize that even if some nonequilibrium problems can be solved there may be phenomena that simply cannot be treated in detail by any possible theory. As Maher points out, the extrapolation from the laws of microscopic physics to the laws of biology might have to pass through regions in which chaotic behavior erects impenetrable barriers. It may in fact be that the hope of truly accurate weather prediction must inevitably be shattered on the shores of chaos.

A final question considered by Maher is whether there are relations among diverse phenomena whose only obvious connection is that they all appear when a certain threshold is passed. For example, once systems grow to volumes greater than about a cubic micron, the statistical second law of thermodynamics becomes inescapable, gravitational effects become important, and quantum fluctuations become unimportant. Maher points out that the tendency to claim deep connections among such phenomena can sometimes be fruitful, but sometimes harmful. Again, a resolution of this question is frustrated by the failure to achieve a satisfactory statistical theory of nonequilibrium phenomena.

Another problem on which a successful statistical theory of nonequilibrium phenomena might shed light is that of "the arrow of time." The second law of thermodynamics tells us that the "correct" direction of time is that in which the total entropy of a closed system increases. Philosophers and scientists have written many tomes discussing the subtleties inherent in that brief statement, but it seems clear that the evolution of systems toward equilibrium is at the heart of the problem. Frank Arntzenius addresses the failure of classical electromagnetic theory to account for the arrow of time in the next paper of this volume.

"The Classical Failure to Account for
Electromagnetic Arrows of Time"
Frank Arntzenius

The basic puzzle connected with the arrow-of-time problem is that apparently irreversible phenomena are described by theories based on time-reversible equations. The standard solution offered to this problem is that the observed one-way flow of time arises from the statistics of large numbers of particles: The evolution is from less probable states, which arise as a matter of initial conditions, to more probable ones. Frank Arntzenius argues that such an explanation cannot be accommodated within the framework of classical electromagnetic theory: The theory cannot give a satisfactory account of the irreversible electromagnetic phenomena that are observed.

One such phenomenon is the irreversible evolution of radiation in a cavity toward the equilibrium distribution now associated with the name of Planck. Arntzenius traces the history of the failed attempts to provide an account of this distribution within the framework of classical theory, focusing particularly on that of Ritz. Einstein almost immediately refuted Ritz's arguments, and it is now a commonplace that it was just this phenomenon that gave birth to the modern quantum ideas.

Arntzenius next discusses the time-irreversible phenomenon of radiation damping. Lorentz provided a classical theory of radiation damping, but it was far from satisfactory. Perhaps the most striking of the difficulties mentioned by Arntzenius is that, according to the classical theory of electromagnetism with radiation damping, atoms would not be stable: Electrons would collapse onto nuclei, and the world as we know it could not exist. But even if such difficulties were not present, Lorentz's derivation of the time-irreversible damping law assumes the retarded nature of electromagnetic waves. Thus this would only explain irreversibility if there were some way of justifying nature's preference for the retarded solutions of the classical electromagnetic equations. Arntzenius examines attempts by Popper and others to do this, and finds them wanting.

Wheeler and Feynman have made what is perhaps the last attempt at a classical treatment of radiation damping. They abandon the assumption of retarded radiation in favor of a time-symmetric theory incorporating both advanced and retarded radiation. They consider a single charged particle surrounded by a perfectly absorbing medium, which emits both retarded and advanced radiation that reaches the particle at a given moment of acceleration. They appeal to statistical considerations to justify their assumption that the retarded radiation, but not the advanced radiation, yields a zero net force on the particle. Arntzenius argues, however, that their appeal to statistical mechanics is, at best, question begging. But in any case, problems such as the classically predicted collapse of atoms would remain.

In his final section, Arntzenius discusses electromagnetic arrows of time within the framework of quantum electrodynamics, and explains why he sees hope that they can be satisfactorily accounted for in a uniform statistical manner within that framework.

"The Reorientation of Neoclassical Consumer Theory"
Edward J. Green and Keith A. Moss

In this paper the authors, Edward J. Green and Keith A. Moss, argue that although the theory of value of J. R. Hicks and the equivalent revealed preference theory of P. Samuelson (the Hicks-Samuelson program) were in some sense failures, it was not the case, as is sometimes claimed, that this program did not conform to any coherent model of scientific practice, nor was it the result of a simpleminded methodological model such as naive verificationism. Rather, the program was a reasonable attempt to develop a conceptually satisfying account of consumers' decisions.

The early economists who modeled consumer behavior as expected utility maximization shared the following assumption with the early utilitarians: Quantities of utility can be compared cardinally within an individual, and perhaps across individuals as well. This concept of utility was then replaced by a weaker one which admitted only ordinal intrapersonal comparisons. It was shown that a formal theory in which an ordinal ranking of individual "consumption bundles" is taken to represent the consumer's preferences is basically equivalent to an ordinal utility theory. It was also shown that one could axiomatize the expected utility theory of decision under uncertainty within this ordinalist framework.

Economists operating within this conceptual framework have met with some success, both in the development of theory and in application of the theory. In what ways has the program failed?

There are several answers. For one thing, the application of these methods to actual data has been much less extensive than some of these researchers expected, especially in light of the vast improvement in computing technology. For another thing, the aggregate demand by all consumers for a set of commodities may not satisfy the theory's formal requirements even if the demand function of each individual does. This matters because individuals conform to the predictions of the theory too imperfectly to confirm it, and one must look to aggregate demand to have any real hope of testability. However, other empirical theories have been successful in accounting for phenomena at the aggregate demand level—theories which can be tested with aggregate data based on the behavior of financial markets, and which require no individual budget data (e.g., Milton Friedman's "permanent income theory").

The failure of the Hicks-Samuelson program thus seems to be due not only to its internal problems but also to the historical accident of the simultaneous success of alternatives. Green and Moss caution us that the fortune of the Hicks-Samuelson model might rise again. Some of the problems of the Hicks-Samuelson model might still be solved, and the other theories falter.

SECTION II—HISTORY: THE ROLE OF FAILURE IN EARLY SCIENTIFIC DEVELOPMENT

"Failure and Expertise in the Ancient Conception of an Art"

James Allen

James Allen argues that important developments in the ancient notions of artistic knowledge and artistic expertise were stimulated by reflections upon the implications of failure by practitioners. Allen argues further that one important effect of these developments was a rudimentary notion of probabilistic knowledge.

To put matters a bit simply, a view commonly held in the ancient world was that to be an artist is to have the knowledge to be able to bring about the object of one's art, invariably, without fail. So if medicine, navigation, and rhetoric were all arts, it would follow that true artists would never fail at these endeavors. But it was the common experience that people who seem to be artists sometimes fail. Therefore, these practitioners fall short of being artists, or what they are practicing are not arts, or this common conception of an art must be revised.

Allen explains that at a fairly early stage a distinction was recognized between arts in which failure was permitted, such as medicine, and arts where it was not, such as making shoes. A physician can sometimes fail to cure a patient and still be practicing the art of medicine, but a shoemaker who failed in an attempt to make a pair of shoes would not have mastered the art of shoemaking. Arts in which failures were permitted were called "stochastic arts," and arts where failure was not permitted were called the "nonstochastic arts."

Practitioners of a stochastic art could have complete knowledge of their art in some sense, and practice their art correctly, and yet fail to bring about results which are the proper end of their art. Some account had to be given of how this relationship between knowledge and the success of practice was possible.

Several proposals were offered. In an Aristotelian tradition it was suggested that the practitioner of a stochastic art can have complete knowledge of the relevant principles and yet sometimes not achieve the desired ends in practice because extraneous factors may intervene unexpectedly and prevent the artist's methods from working properly. This was consistent with the Aristotelian idea that it is in the very nature of some principles of nature to be true only "for the most part."

Another suggestion, in a rationalist spirit, was that true principles of, say, medicine are necessarily true. But the body of such principles knowable by a physician is abstract and insufficiently detailed to prescribe diagnosis or therapy amid the many idiosyncratic details of a particular case.

It is not clear how this doctrine constitutes an account of the kind needed, and indeed an empiricist school of thought did not buy it. These empiricists saw the principles constitutive of medical knowledge as themselves inherently "stochastic." The physician knows how to find signs in nature but the presence of these signs correlates only to some extent with the presence of the thing signified. Here we have the germ of a frequency-based conception of probability.

"Scholasticism and the Philosophy of Mind: The Failure of Aristotelian Psychology" Peter King

In this paper, Peter King discusses the failure of a "research program," in a broad sense of the term. Here a research program is a nexus of common assumptions, common methods of exploration and validation, common views about promising lines of research, and a common set of terms for scientific debate. Aristotelian psychology was such a program. Its collapse in the Middle Ages was due to its inability to solve what King calls the problem of "transduction."

The kind of transducer at issue here is a psychological mechanism which maps physical input coming in through the senses onto output which is intellectual in the sense that it is language-like, and manages to do this even though the input does not itself have syntactic or semantic structure. The problem of transduction is to find such a psychological mechanism to mediate between the sense and the intellect.

This was a pressing problem for the Scholastics since most of them accepted the following principles: First, the distinction between sensing and knowing was a distinction in kind based on differences between the faculties of sense and intellect. Second, the understanding may be characterized linguistically so that concepts are thought of as mental words. Third, the intellect is initially a *tabula non scripta* and so the mental vocabulary must be acquired. It followed from these assumptions that there must be a transductive mechanism.

One version of the story goes like this: Knowledge of wolves, that is, knowing what a wolf is, consists in the form *wolf* being transferred from sensory input to the "potential intellect." But then the intellect actually has this form—what was only potentially wolf has become really, formally, wolf. Something in itself active must have been involved to account for this.

And since the state of the intellect being actually wolf is a conceptual "linguistic" state (the possession of a mental "word"), the active thing that is responsible for this process must be a transducer.

The transducer is called the "agent intellect." It works by abstracting the universal form from the idiosyncratic material in which it is instantiated, and then redepositing it in the potential intellect. This was St. Thomas's version; there were others. The problem here is that the agent intellect must abstract out the form—say, *wolf*, for instance—not only from particular matter but from other universal forms—say, irrationality—with which it is coinstantiated. In order for the agent intellect to manage this it must apply symbolic analysis to its data. It is not acting as a transducer after all.

The other versions of the story met similar fates. Despite much scurrying about in an attempt to get around the problem, nobody really succeeded until Descartes merged sensation and intellection in a single mind.

SECTION III—SOCIAL SCIENTIFIC PERSPECTIVES

"Converging Failures: Science Policy, Historiography
and Social Theory of Early Molecular Biology"
Pnina G. Abir-Am

Although, as noted earlier, philosophers and scientists alike have slighted the study of scientific failure, a perusal of the references in Pnina G. Abir-Am's paper shows that social scientists have not done so. This paper carries such investigations further by using an historical case study in early molecular biology as a vehicle for an examination of the convergence of notions of failure in science policy, historiography, and social theory.

In the 1930s, an interdisciplinary team of scientists at the University of Cambridge was working on physicochemical embryology. Both the topic and the interdisciplinary approach were of interest to the Rockefeller Foundation, which awarded several grants to the Cambridge team. Before deciding on whether or not to consolidate and enlarge their support of such work with a long-term grant, the Foundation's officers attempted to assess advice from the scientific community on the project. Their assessment eventually led to a decision not to grant any further support to the project, which came to be considered a failure.

Abir-Am argues that it was not the science that was a failure; rather, it was a cluster of failures in science policy that led to termination of the project. These failures can be grouped into two broad categories: failures in the way advice was solicited and failures in the way the advice was processed once it was received. Among the failures in the first category were failure to select relevant, rather than merely easily accessible, advisors; failure to give appropriate guidelines to the advisors; and confusion of scientific questions with ones of policy. Among the processing failures were failure to recognize the contextual constraints of the advice received; the misconstrual of minor reservations as being of more serious import; and failure to differentiate between casually given advice, which carried little importance, and well argued, highly relevant advice.

Abir-Am notes that some historians as well as some scientists have reconstructed this history as one of scientific failure of the research program. After examining some of the reasons for this, she turns to an examination of how to make this translation of a policy failure into a scientific one intelligible. Her approach is in terms of a model of scientific change in which transitions occur among three levels: the microlevel of individual investigators, the mesolevel of a collaborative team, and the macrolevel of "permanent" institutional existence. The transition from the micro- to the mesolevel had been successfully accomplished by the Cambridge team, but the next transition failed to occur. The mesolevel is, in Abir-Am's terminology, a metastable state, one that is of short duration and sensitive to disturbances. Such disturbances can cause the metastable state either to revert back to the stable microlevel of individual action or to progress to the stable macrolevel of social structure. Thus Abir-Am finds a symmetry between failure of scientific change to take place and success. They are both transitions from the metastable state, but in opposite directions. In the case at hand, failure of scientific policy was the disturbance that caused a transition back to the microlevel, and it was this transition that caused the scientific program to be seen as a failure.

"The Determinants of a Scientist's Choice of Research Projects" Arthus M. Diamond, Jr.

In the aftermath of Kuhn, many philosophers and sociologists of science have explored the question as to what extent the direction that science takes depends on factors "internal" to scientific theories—such as their explanatory power—and to what extent it depends on "external" factors—such as the prejudice, the ideology of the practitioners, their geographical location, age, or economic interests. Although there has been much speculation as to the relative importance of various external factors, little has been done to try to quantify their relative contributions. Arthur M. Diamond, Jr. attempts to quantitatively measure the contribution of some external factors to the choice of a research program. Diamond's target is the brief episode in the history of science when some scientists chose to write on "polywater."

"Polywater" was thought to be a new form of water that had a lower freezing point and a higher boiling point than ordinary water. In the West, a surge of articles on the stuff began to appear in 1969. Four years later everyone agreed that there was no polywater, and the original observations were due to impure samples. Diamond collected biographical information on many of the scientists who wrote on polywater. He chose to study 100 scientists who wrote on polywater and 100 scientists who did not but could have, given their areas of expertise. Then, among scientists who chose to write on polywater, he classified their articles as pro, con, or neutral.

It has been claimed that older people will cling to old theories. Max Planck famously said, "[A] new scientific truth does not triumph by convincing its opponents and making them see the light, but rather because its opponents eventually die, and a new generation grows up that is familiar with it." But Diamond discovered that age (measured as years since Ph.D.) was not a statistically relevant factor. Also irrelevant were various other factors that might contribute to risk aversion among older scientists, such as the income or number of children they had. In fact, scientists with tenure were more likely to undertake risky projects.

Some other factors that seem to contribute to choice or research program were the availability of research funds, the interest of journal editors, and proximity to other researchers interested in the project.

Certain factors that Diamond did not test could be subjects for future research—for example, whether the project appears to have practical applications or whether it looks like fun.

Tamara Horowitz

Allen I. Janis

I. METHODOLOGY:
THE ROLE OF FAILURE IN
MODERN SCIENTIFIC DEVELOPMENT

1. THE VALUE OF
SCIENTIFIC FAILURE*

Allen I. Janis

Einstein has been quoted as saying, "Science can progress on the basis of error as long as it is not trivial" (quoted in Koshland 1988, 1261). My purpose here is to give a brief discussion of the scientific value that can be derived from three distinct types of scientific failure.

The type of failure that is discussed most often is when a theory is found to be incorrect (or at least incomplete) as a result of newly discovered observations. This type of failure is at the basis of much of scientific progress, and examples abound. A relatively recent example is provided by advances in the study of superconductivity. For almost three decades, it was generally accepted[1] that superconductivity was best described by the so-called BCS theory of 1957, for which Bardeen, Cooper, and Schrieffer received the Nobel Prize in 1972. In this theory, superconductivity results from the pairing of electrons in solids through their interaction with lattice vibrations (in technical terms, through the exchange of virtual phonons). Although the theory itself does not give any clear prediction of an upper bound to the critical temperature, T_c, above which superconductivity will not take place, it was generally thought that the BCS mechanism could not lead to superconductivity at a temperature as high as that of the boiling point of liquid nitrogen (77 K).

Early in 1987, reports came out of finding superconductivity at temperatures above 90 K. This created such excitement in the scientific community that a panel discussion of these results shortly thereafter, at a meeting of the American Physical Society, became what was described in news accounts as "a Woodstock for physics." Only a few minutes after the doors of the meeting room opened, 45 minutes before the discussion's scheduled starting time of 7:30 P.M., all 1140 seats were filled by those who had been waiting outside for more than an hour. Hundreds more stood in the aisles for hours to hear the discussion, and many more watched on TV monitors placed outside. More than a hundred were still present when the session was finally brought to a close at 3:15 A.M., and some even remained until the room was reclaimed by the hotel for other purposes at six in the morning. I recount these events to indicate the intense interest that the new observations generated. A news account of these events, including a somewhat technical discussion of the experimental and theoretical background, may be found in Khurana 1987.

It was not long before theorists came up with plausible mechanisms, different from that of BCS, to explain these new results, and the theory of superconductivity is now enhanced considerably beyond its state before these discoveries. Thus the discovery of results that the BCS theory failed to explain led to advances in the theoretical understanding of superconducting mechanisms, advances that undoubtedly would have come much more slowly, if at all, than would have been the case had observations of superconductivity continued to fit nicely into the BCS framework.[2]

As I indicated at the outset, this type of failure is almost prototypical of the way in which science proceeds. Progress in engineering proceeds in similar ways, as is described by Petroski (1985) in a book subtitled *The Role of Failure in Successful Design*. In Petroski's view, "the colossal disasters that do occur are ultimately failures of design, but the lessons learned from those disasters can do more to advance engineering knowledge than all the successful machines and structures in the world" (*ibid.*, xii).

The type of failure I have been describing is that of a theory that had been generally thought of as correct and complete (at least to the extent that theories are ever thought of as correct and complete) until new evidence came along to show that it was either in need of modification or only of limited applicability. Some failed theories, on the other hand, were never generally accepted or were viewed as no more than one of a number of possibilities; and then when further evidence came along they were discarded even by most of their former supporters. Even such failures can be of scientific value if, in the course of solving problems raised by these theories, their proponents achieve significant advances in scientific understanding that turn out to be independent of the success or failure of the theories that had provided the motivation for studying those problems.

The so-called steady state theory of cosmology provides an excellent example of this type of fruitful failure. This theory, proposed in 1948 in one form by Bondi and Gold and in a somewhat different form by Hoyle, differs from the cosmological theory provided by general relativity in that the latter has the universe evolving from a "big bang" (a singularity in the geometrical structure of spacetime) at some finite time in the past, whereas in the former the universe has existed forever in more or less its present form. Since the expansion of the universe, which is accepted as being a true feature of the universe in both theories, would otherwise necessarily lower the average density of matter in the universe, the steady state theory has matter being continually created throughout the universe (for a description of the steady state theory, see Bondi 1960, chap. 12). Although both the big bang and steady state theories were viewed by at least a significant fraction of the scientific community as viable possibilities for a good many years, ultimately most scientists agreed that the preponderance of evidence weighed against the steady state theory (see, for example, Raychaudhuri 1979, 3).

In order to understand why the steady state theory was nevertheless of considerable scientific value, it is necessary to look at the problem of formation of the elements (for a summary of the history of this problem and its current status, see Narlikar 1988, 81-96, or, for a somewhat more technical discussion, Raychaudhuri 1979, Chap. 7). The earliest treatments of this problem, by Gamow and his co-workers, supposed that the elements were formed, out of particles and radiation that existed shortly after the big bang, through nuclear reactions that could take place at the temperatures and densities that existed then. Proponents of the steady state theory could not, of course, accept this as the correct explanation. Since it was supposed in their history that the newly created matter was hydrogen (or neutrons, whose decay would produce the constituents of hydrogen), they had to find other ways of creating the remaining elements. Hoyle and others found that nuclear processes in stars (including those that take place during violent events such as the explosion of a supernova) could produce heavier elements starting just from hydrogen. Their success in working out the details of these processes produced such a convincing scenario that now even proponents of the big bang theory, having taken a closer look at just what nuclear processes were possible in the conditions obtaining in the early universe, acknowledge that most of the elements were indeed "cooked" in stellar furnaces. Although, in order to explain the observed abundances of the elements, it is necessary to suppose that some of the formation of lighter elements took place in the primordial "soup" existing not long after the big bang, all the rest is now generally believed to be the result of stellar processes. In 1983, Fowler won the Nobel prize for his work on these processes; much of this work was done in collaboration with Hoyle, one of the founders of the steady state theory.

Thus, although the steady state theory is now generally thought of as a failure, it spawned our present, generally accepted view of the creation of the heavy elements. This understanding would undoubtedly have come along in due course even if there had been no steady state theory, but it seems clear that this theory considerably hastened our reaching this understanding.

The third type of failure that I wish to discuss is the failure to take account of all of the relevant experimental data when comparing a theory with experiment. Instances of this type of failure are not uncommon and range over a variety of motivations. In some cases, it is explicitly acknowledged that certain aspects of a phenomenon under investigation are not explained by the theory, but it is felt that the theory nevertheless explains enough previously puzzling aspects that it seems worth pursuing. On the other hand, in some cases, data have been deliberately suppressed in order to make a theory look better than it really is. Although instances of the latter type are not likely to be of much scientific value, it can sometimes be of considerable benefit to ignore certain parts of the seemingly relevant data. An interesting example is provided by the early history of the quantum theory of solids.

The very first paper to apply the still new quantum ideas to understanding the nature of solids was Einstein's treatment of specific heat, which was submitted for publication late in 1906 (for a description of Einstein's theory of specific heat as well as the associated history of the problem see Pais 1982, chap. 20). The success of this theory in explaining the behavior of the specific heat of solids at low temperatures is generally considered to be one of the major milestones in the early history of quantum theory, and it led to the further use of quantum ideas in explaining other properties of solids. Einstein made a number of simplifying assumptions in the development of this theory, which he realized would limit the accuracy of his calculations; one of these was that all of the atoms, which were located at the lattice points of a three-dimensional crystal lattice, vibrated with the same frequency. The theory he developed had one adjustable parameter in it, which was this vibration frequency.

At that time there were two sets of relevant data on specific heats at low temperature; one, due to Weber, was in the range from 1000 °C to –100 °C, and the other, due to Dewar, was in the much lower range of 20 K to 85 K (note that –100 °C is 173 K). It seems that only Weber's data were known to Einstein, and he used the adjustable parameter in his theory to obtain a best fit of his predicted values to those data. The resulting fit, which Einstein showed on a graph in his paper,[3] turned out to be excellent over the entire range of Weber's data.

Had Einstein considered the combined data of Weber and Dewar, however, he would not have had such great success. Einstein's predicted behavior for the specific heat falls off exponentially at low temperatures, whereas the correct behavior, as first found by Debye, is that the specific heat falls off as the third power of the Kelvin temperature. The difference would have been noticeable had Einstein tried to accommodate Dewar's data as well. One can only speculate as to whether he would have thought he was on the right track had he known of Dewar's data. The fact that he took the rare (for him) step of including a comparison graph in his paper shows that he attached considerable importance to the goodness of fit. It is also known that the lack of agreement of earlier versions of general relativity with the precession of the perihelion of Mercury was one of the factors that caused him to be dissatisfied with those earlier theories (see, for example, Pais 1982, 250). Perhaps he would have tried to improve the approximations in his theory of specific heat, but it is also known that later, after learning from Nernst's improved measurements of specific heat that his theory was inadequate, Einstein explored some modifications of his assumptions without complete success.

It is, of course, impossible to know what Einstein would have done had he been aware of Dewar's data. Perhaps he would have decided that the qualitative success of his theory in showing that the specific heat decreased at low temperature was sufficient to warrant publication of his ideas in spite of the fact that the low-temperature behavior of his theory was not in quantitative agreement with all of the data. In hindsight, at least, that would clearly seem to have been the correct course. In any case, however, it seems clear that if the disagreement with Dewar's data had not been ignored, the significant progress in the development of quantum theory that resulted from Einstein's study of the specific heat of solids would at least have been considerably delayed.

These examples illustrate the force of the quotation that opened this brief essay. Each shows a theory that, sooner or later, failed to hold up to the weight of evidence. But the errors were not trivial ones: The mistaken ideas had enough truth in them that their pursuit led to substantial scientific progress.

University of Pittsburgh

Notes

*I am grateful for Gerald Massey's helpful comments on an earlier version of this paper.

1. Some, of course, questioned whether there might not be forms of superconductivity not properly described by the BCS theory. In hindsight, they were clearly correct. Nevertheless, the remarkable interest shown in the discovery of high-T_C superconductors attests to the fact that their discovery was neither generally expected nor in accord with any generally accepted theory. A full account of the recent history of superconductivity is beyond the scope and intent of this present discussion.

2. Because of the pejorative implications of the word "failure," some people might object to my use of it in connection with the BCS theory. I do not, in fact, see the later developments as diminishing the importance of the BCS theory. Had the BCS theory not already been in existence to serve as a model, the newer theories would almost certainly not have been developed so rapidly. I use "failure" only to indicate the fact that it failed to explain the new observations.

3. This was, in fact, one of the rare occasions when a paper of Einstein's contained a graph that compared theory with experiment. Pais remarks that he knows of only two other instances in all of Einstein's works; see Pais (1982, 389).

References

Bondi, H. (1960), *Cosmology*. 2d ed. London: Cambridge University Press.

Khurana, A. (1987), "Superconductivity Seen above the Boiling Point of Nitrogen," *Physics Today 40* (no. 4): 17-23.

Koshland, Jr., D. E. (1988), "Random Samples" (editorial), *Science 240*: 1261.

Narlikar, J. V. (1988), *The Primeval Universe*. Oxford: Oxford University Press.

Pais, A. (1982), *'Subtle is the Lord...': The Science and the Life of Albert Einstein*. Oxford: Oxford University Press.

Petroski, H. (1985), *To Engineer is Human: The Role of Failure in Successful Design*. New York: St. Martin's Press.

Raychaudhuri, A. K. (1979), *Theoretical Cosmology*. Oxford: Oxford University Press.

2. THE FRUSTRATED STATE OF NONEQUILIBRIUM STATISTICAL PHYSICS

James V. Maher

There are obviously many ways to define scientific failure, especially since there is no existing scientific theory which is not subject to at least some criticism. Rather than worry too much about finding a rigorous definition of failure, let me take the "I know it when I see it" approach, and say that, for present purposes, a scientific failure is present when there is a significant lack of success over a protracted period of time in solving a problem which the scientific community agrees is a significant problem. Almost every word of this loose definition invites some discussion, and much of such discussion and clarification of terms will be woven into the text below. I propose to take as a conversation piece the current state of nonequilibrium thermal and statistical physics. I will use this field of modern physical research to illustrate some interesting features of a failure and to launch a speculation on which of these features might be universal in situations of failure. In choosing nonequilibrium statistical physics as an example of failure, I will encounter no argument from either the scientific or philosophical community; it will become clear in the discussion below that this case easily fits the definition above and would probably fit any reasonable modification of the definition.

Nonequilibrium statistical mechanics has been under investigation for roughly a century now, and its macroscopic sister science, nonequilibrium thermodynamics, is almost two centuries old. The entire catalog of successes in the field consists of a lamentably small list of special, simple cases, with no successful general theory. Because the field has so few successes, it is not discussed much. Until very recently, students were not exposed to it, and even now, it is typically discussed only in special-topics courses at the graduate level (for a comprehensive discussion of this field, see Kubo and others 1985). This may reflect a typical approach to scientific failure; our public discussions dwell on our successes. Let us now examine some of the scientific issues in statistical physics, and as we do so, let us consider the following questions: (1) Can we identify the source of the failure? (2) Has much good come of the apparently futile work of the last two centuries? and (3) Can we say anything in general about how scientists behave when things are not going well?

As a background for discussing nonequilibrium statistical physics, let me give a brief summary of some of the issues in equilibrium statistical mechanics.[1] The equilibrium theory is a great success by my definition above since the scientific community is virtually unanimous in declaring it to be so. There are some very fundamental questions about the connections between microscopic workings of classical and quantum mechanics and assumptions of randomness which trouble physicists a little and philosophers a lot (for a discussion of the philosophical issues, see Earman 1986, esp. chaps. 8 and 9 and references contained therein.) However, I dodged this problem by letting the scientists decide on which scientific fields have failed, and even the philosophers can temporarily put aside their worries about the equilibrium theory and agree that the nonequilibrium theory is in enormously worse shape.

The equilibrium theory applies only to systems which are in an equilibrium state, and an equilibrium state is defined to be one in which the macroscopic observables no longer depend in any way on the history of the system and are no longer changing with time. This is a very useful approximation to a reasonably large number of attainable physical situations, but it is only an approximation. Nothing is at equilibrium if you are too fussy in applying the definition. For example, a bar of iron can have a rather well defined temperature if one controls its environment with any modest effort, but if one looks really closely, the iron is slowly rusting away on time scales of years, and its exact state is technically a function of its history. In practice, we do our experiments fast in comparison to any time scale on which the iron is changing, and we experience great success in explaining its properties with equilibrium theory.

The goal of statistical theory, equilibrium or non, is to predict macroscopically observable properties of large systems in terms of the systems' underlying microscopic properties. For example, one of the early triumphs of equilibrium statistical mechanics was to recover all of the known thermodynamic properties of the ideal gas by treating a mole of ideal gas as Avogadro's number of noninteracting atoms which share a container. (This success was achieved mostly by appeal to classical mechanics, but some help was needed from quantum mechanics to get details like the specific heat straight.) The most powerful tool in relating the equilibrium microscopic and macroscopic worlds has been the ensemble. An ensemble is an imaginary array of systems which are macroscopically identical; in this array every possible microscopic configuration of constituent particles is represented with equal weight to that of any other individual possibility. The theoretical physicist constructs the ensemble by calculating the array of possible microscopic variations which are consistent with whatever macroscopic constraints have been placed on the system in question. Then any observable property can in principle be calculated by averaging the expected results of the measurement for each of the elements of the ensemble. The assumption is always made that each possible microscopic configuration is as likely as any other, and with this approach there has been great success in predicting the thermodynamics of a wide variety of simple systems.

The greatest hope for extending the success of equilibrium statistical mechanics to nonequilibrium situations lies in exploiting the insights provided by the equilibrium theory as it successfully predicts the properties of the new equilibrium a system will find after it is forced out of an old equilibrium state by having a constraint relaxed. It is well known that a system will spontaneously increase its entropy, and this result of classical thermodynamics can be understood easily if we define the entropy to be the logarithm of the number of microscopic states accessible to the system. To illustrate the implications of this definition of entropy, let us consider a well-insulated container enclosing N molecules of an ideal gas; no energy can flow into or out of this container. As our macroscopic observable, let us measure the center of mass of the gas. For an initial equilibrium state, imagine that the container is divided in half by an impenetrable barrier and that all gas molecules are on the left side of the barrier. The ensemble appropriate to this state is the set of all possible configurations of energy E which keep all N molecules on the left side of the container. Almost all of the very large number of arrangements of N molecules will put the center of mass of the system very near the center of the left half of the container. (In fact, the distribution of centers of mass of the ensemble elements will be a Gaussian or normal distribution whose maximum lies at the center of the accessible half of the container and whose standard deviation is a very small fraction of the size of the container, about a part in 10^{+12} if N is Avogadro's number.) There will indeed be configurations whose center of mass is observably displaced from the center, but there will not be many of these, and if one makes the usual assumption that each configuration is equally likely, then a system which changes configurations every 10^{-12} seconds would have a negligible chance of visiting one of these anomalous configurations even once in a time span equal to the present age of the universe.

Now imagine removing the partition from the center of the container. The system still has energy E and number of molecules N, but it is now free to sample arrangements which fill all of the container. The system now has an enormously larger number of arrangements available to it, and almost all of these have a center of mass near the center of the container; the old equilibrium arrangements are all still allowed, but they will now be incredibly improbable once the system forgets its recent history and settles down to its new equilibrium. The figure shows the number of configurations, P, in the new equilibrium ensemble as a function of the center of mass, x. Even on the logarithmic scale used in the figure, the number of configurations in the original equilibrium ensemble is not visible, even though these are themselves a rather large number of configurations whose average value lies at position x_0. Thus we do not expect the system ever to show us a configuration with center of mass zero after we attain the new equilibrium; this is essentially the same statement as the classical thermodynamic result that the system entropy increases when the partition is removed and the gas will never spontaneously go back into the left half of the container.

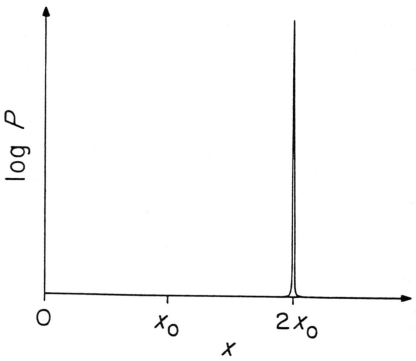

Figure.

The discussion above allows us to say something about spontaneous fluctuations and it also motivates the only real hold we have on nonequilibrium phenomena. It tells us that there are spontaneous fluctuations (something not contained in classical thermodynamics) but that these are virtually guaranteed to be small. This guarantee is so reliable that when we do see a system with a value of x which is far from the equilibrium expectation, we assume that someone has been imposing constraints on the system until very recently and that the system will rapidly leave x never to return. This leaves us with the prejudice that there is no intrinsic difference between a system undergoing a small spontaneous fluctuation and a system which has been pushed away from equilibrium by the same amount. The only successes of nonequilibrium theory are based on positing this identity between the response of a system to a small disturbance and the response of the same system to a small spontaneous fluctuation. If a disturbance is large in comparison with expected fluctuations, we have no demonstrable point of contact with equilibrium theory, and in general for such a case, we expect an overwhelming increase in mathematical complexity. Now that we are oriented with the above discussion of equilibrium spontaneous fluctuations, let us return to discuss the nature and possible sources of failure in the nonequilibrium cases.

A few examples of nonequilibrium systems will illustrate the range of the phenomena, their technological importance and their fundamental importance for understanding connections between microscopic physics and other branches of the physical and biological sciences: (1) Turbulent systems are intrinsically far from equilibrium. If we had a good, manageable theory of turbulence, we could predict the weather and learn how to mix fuels more efficiently with their oxidants. (We would also have a much better idea of how microscopic forces couple up to form macroscopic mechanical forces, both dissipative and non.) (2) Many of the interesting materials in our world, from naturally occurring glasses to some of the most exotic materials needed for our most advanced technologies, are highly disordered systems trapped far from equilibrium in very long-lived metastable traps. (3) Many of the fascinating physical patterns one sees in nature, such as snowflakes, dendrites in alloys and the nervous system, and viscous fingers in oil fields, result from processes which take the systems far from equilibrium during the most crucial stages of pattern formation. (4) Most important of all, biological systems are obviously far from equilibrium during the entire time that the systems could be said to be living. (The equilibrium state for a frog in a closed container would not be reached until, after death, all the frog's tissue had decayed away to much simpler, stable chemical species.) The reductionist argument that all laws of biology could in principle be recovered from careful application of the laws of microscopic physics cannot be made even in principle unless one asserts that we have an acceptable understanding of nonequilibrium statistical physics.

We cannot realistically calculate the properties of any of the phenomena listed in the preceding paragraph. This failure exists both for macroscopic and for microscopic systems, and it is unambiguous if one requires quantitative results of our theories. Let us now examine some of the responses of the physics community to this problem. The first response, and one that was nearly unanimous before about 1970, asserts that there is no failure, rather there is an excessively messy problem of detail, but no problem of principle. While it may be true that the equations one would write down from microscopic physics should in principle work for large systems, these equations would constitute an enormous set of coupled, possibly nonlinear, partial differential equations. Any attempt to truncate the unworkably large number of equations would inevitably lead to introducing important nonlinearity into the resulting differential equation(s). The nonlinear classical equations which are believed to be relevant to such a program have attainable operating points which are singular; in fact, some such points can be shown to be so pathological that no neighborhood of the point is sufficiently small that it lacks points whose trajectories diverge exponentially from that of the original point—such points are important in the recently active field of chaos research (for a discussion of the mathematical properties of singular points in the chaotic regime, see a text on deterministic chaos; for example, Schuster 1988). When a classical system is operating near such a singular point, its time evolution is indeterminate even though the classical equation it "obeys" is deterministic. Continuous improvement in the experimental control of the operating point would not lead to a deterministic situation; instead, the situation would remain chaotic until the control approached the limits imposed by quantum mechanics, at which point the classical equation would no longer be appropriate. This admittedly sketchy discussion should be sufficient to create dissatisfaction with the argument that the existing theory is in principle sufficient. It is clear that there are important problems on which we cannot make progress because the existing theory is at best unwieldy. In addition, it is highly unsatisfactory to have so much of the connection between our microscopic theories and the theories of the macroscopic world be untested.

Given that most people nowadays agree that there is a problem, what other responses are available? The most popular response is to blame it all on the mathematicians. There is no general theory of nonlinear differential equations, and what we do know of the specific nonlinear equations of greatest physical importance is very sketchy in comparison to our usual expectations (for an example of a good approach to nonlinear differential equations at the advanced undergraduate level, see Boyce 1986, chap. 9). The great successes of physics have arisen with situations governed by linear differential equations where we know uniqueness and stability conditions and can construct ever more complicated solutions by superposing elements in a library of simple solutions. None of these advantages apply to the physically relevant nonlinear equations (or indeed to any but a very few, exceptionally simple toy equations): (1) Lacking a uniqueness condition, even after painfully constructing a detailed solution to the equation, one does not have any confidence that nature would not choose some other equally valid solution; (2) since the sum of two solutions is not in general a valid solution, there is no way to use simple cases to work one's way to more sophisticated solutions; and (3) it is usually more difficult to investigate the stability of a solution to small changes in the initial or boundary conditions.

There has been some progress recently, and this has occasioned the renewed interest in the physics community in some of the long unsettled issues of classical physics.[2] This progress has come in two forms. Increased computational power has allowed the brute force construction of solutions to a few of the simplest of the physically interesting nonlinear problems. Such calculations must be kept in close contact with experimental results because of the uniqueness problem, but they have at least become feasible. In addition to the computational advances, powerful mathematical techniques, some dating back to the beginning of this century, have been combined to provide previously unavailable insight into the character of the possible solutions. Thus we find the recent physics literature overflowing with references to field theoretic (like the renormalization group), topological categorizations (e.g., strange attractors) and especially with the language of infinite set theory (such as fractal or Hausdorf dimension, Cantor set, and Julia set).

By the end of the next decade we will probably know whether or not this flurry of mathematical activity has been successful, but for present purposes it probably illustrates adequately a general feature of the response of the physics community to failure: Having had past success by attacking difficult problems by applying successively more sophisticated mathematics, we are currently throwing all the math we can find at this problem in a frenzy of activity. This frenzy may itself appear chaotic to an outsider, and it certainly involves much less attention to rigorous thinking and the rigorous use of mathematics than would be normal when things seemed under better control, but there is an internal logic and beauty to the enterprise which most of those involved find very stimulating and far more exciting than following a clear path.

Now that we have sketched the argument that says we can blame the whole thing on the mathematicians, let us return to our main question: Is there a failure of the physics? We can now see this question to be asking: Do we have the physics right in these equations which we cannot solve? Obviously we do not know, and this in itself is a failure. We can, however, make a few useful comments.

First, the few mathematical successes have tended to support the argument that the physics is just fine and that all we need is more computational power.

Second, there have been interesting recent interactions between experimenters and simulators which give hope that the physics is on the right track and that we can indeed calculate ever more complicated situations. Simulators are in principle neither theorists nor experimenters. They set up computer models of the physics which is believed to underlie a complicated statistical problem, and then they repeatedly calculate the behavior of the physical system, changing their initial conditions randomly for each calculation. Simulators can frequently achieve much better statistical accuracy than experimenters (depending on how sensitively the outcome of a realization of the experiment depends on small fluctuations in the initial conditions), but they can never be sure that they have included all the important physical terms in their models. When the simulations and the experiments agree, however, one can gain some confidence that all the important underlying physics has been correctly included and that there is no unanticipated richness to the set of possible phenomena. Such concurrences of experiment and simulations have occurred recently in some simple but important and previously intractable problems.

Third, one can hope that we are making progress on some of the large and important nonlinear physics problems. For instance, there is reason to believe that we may be on the verge of describing some disordered materials by adding a little order to models of complete randomness. The models of complete randomness—for example, diffusion limited aggregation (DLA) models of chaotic growth of structures and random field theory models of the microscopic physics of disordered materials—work at present only qualitatively and only for a very restricted class of materials where microscopic disorder is extreme. However, experimenters have been able to demonstrate reasonable agreement with such models under very carefully controlled circumstances (where ironically the control is used to ensure that randomness is maximized). This encourages one to believe that great success will follow once the theories are improved to include a little order. Such improvements would be analogous to the development earlier this century of solid state physics where significant successes followed the introduction of models of imperfections in crystals; nature appears not to make either perfect crystals or perfectly disordered systems.

The discussion above has been rather diffuse because the subject matter is multifaceted and because the physics community has as yet been unable to impose any unified viewpoint on the complex of problems. In other words, it is difficult to give a coherent discussion of an area of failure. As a last step before drawing any conclusions about failure, let us now draw on the various insights presented above and try to articulate the important scientific and philosophical questions which are being blocked by our failure to develop a comprehensive theory of nonequilibrium statistical physics. Three such questions are of particular importance.

The first is the question to which we have returned repeatedly in the discussion above: Do we know all of the physics which is needed, or is our failure more than merely mathematical? It is clear that we do not know and are unlikely to know the answer to this question in the near future. There appears to be progress on some few problems where large scale simulations give a good account of complicated stochastic processes without invoking more than simple models for the underlying microscopic physics. On the other hand, there have been other recent successes, notably the progress in chaos research, which dramatize the richness of the phenomena which can emerge from very small changes in operating conditions of nonlinear systems. One must wonder whether these phenomena will require for their understanding a modification of the rules we would like to use to predict macroscopic phenomena from known microscopic rules.

Even if the rules we want to use to connect known microscopic phenomena to macroscopic behavior will never need to be revised, there are still serious questions regarding the reductionist prejudice which is almost universal in the physics/chemistry (but not the biology) community. That is, are all macroscopic phenomena in principle recoverable purely from a knowledge of microscopic law? One example will suffice to illustrate the way in which resolution of the reductionist issue is blocked by the failure to develop a nonequilibrium statistical physics. Biology concerns itself with organisms which are intrinsically far from equilibrium; in order to reduce a biological specimen to true thermodynamic equilibrium, one must wait an extremely long time for slow natural processes to reduce the tissue to carbon dioxide, water, and such, or else produce some violently nonequilibrium process like a cremation fire. In order to try to learn the laws of biology from the laws of microscopic physics, one must learn to extrapolate equilibrium statistical physics a very, very long way into the nonequilibrium regime. The extrapolation could well involve passing through parameter regions which are dense with chaotic singular points, in which case modern chaos theory could well be saying that the only way to discover the laws of biology, in principle as well as in practice, is to learn them just the way the biologists have always done so.

The final major question whose resolution is being blocked has to do with relations among the peculiar phenomena which seem to appear once a system at normal earthly density reaches a size greater than or equal to a cubic micron (see Penrose 1986, 1987). At roughly this size the following three effects appear: (1) The second law of thermodynamics becomes inescapable, forcing one to accept an apparent arrow of time or irreversibility; (2) gravitational terms from all atoms in the object become sufficiently large for any one atom that any quantum mechanical calculation would need to explicitly include gravity (something which we do not at present know how to do); (3) quantum fluctuations become small enough that objects this large become suitable detectors in the sense of the quantum mechanical measurement problem. Some would like to argue that these three effects are deeply related, others that there is no good reason to believe that they are related in any deeper sense than that they share a critical length scale. The resolution of the debate is frustrated by the lack of a statistical theory of time-dependent phenomena.

Having considered the state of this field of physics and its implications for questions of general interest, I will conclude our discussion with two remarks. The first is the simple and inescapable conclusion that we have failed to connect the microscopic with the macroscopic in a truly convincing way. Statistical nonlinear dynamics is too much for us, and this is a failure which bothers both physicists and philosophers. The second remark is more speculative: Let me attempt to generalize the observations of this one example of nonequilibrium statistical physics to situations of failure in general. A scientific failure like this becomes a bottleneck and some features of a bottleneck appear to be universal. When science encounters a bottleneck, five characteristic features result: (1) At first, the community tries to ignore the problem, as was the case for our example through most of the last seventy years; (2) problems keep piling up; successful related areas keep posing more questions which highlight the failure by requiring its resolution for their resolution; (3) unable to solve or ignore the problem, the scientists resort at least partially to name calling (as in blaming it all on the mathematicians); (4) the problem area becomes a fertile ground for trying out every novel technique which appears in any field; in particular, any available mathematics, no matter how exotic, is brought to bear on the problem, with the question of why it might be relevant deferred until after some success is demonstrated; and finally, (5) there is some tendency to claim deep connections among all of the questions raised by related areas. This tendency is sometimes fruitful and sometimes harmful.

University of Pittsburgh

Notes

1. There are many excellent textbooks on equilibrium statistical mechanics. One example is Huang (1987).

2. The literature in this field is extensive but diffuse. For a good, pedagogical collection of papers spanning experiment, simulation and theory, see Stanley and Ostrowsky (1989).

References

Boyce, W. E. and DiPrima, R. C. (1986), *Elementary Differential Equations*. New York: Wiley.

Earman, J. (1986), *A Primer on Determinism*. Dordrecht: Reidel.

Huang, K. (1987), *Statistical Mechanics*. 2d ed. New York: Wiley.

Kubo, R; Toda, M. and Hashitsume, N. (1985), *Springer Series in Solid State Sciences*. Vol. 31, *Statistical Physics II: Nonequilibrium Statistical Mechanics*. 2d ed. Berlin: Springer-Verlag.

Penrose, R. (1986), "Gravity and State-Vector Reduction," in R. Penrose and C. J. Isham (eds.), *Quantum Concepts in Space and Time*. Oxford: Oxford University Press, pp. 126-46.

————. (1987), "Newton, Quantum Theory and Reality," in S. W. Hawking and W. Israel (eds.), *Three Hundred Years of Gravitation*. Cambridge, England: Cambridge University Press, pp. 17-49.

Schuster, H. G. (1988), *Deterministic Chaos*. 2d ed. Weinheim: VCH.

Stanley, H. E. and Ostrowsky, N. (eds.) (1989), *Random Fluctuations and Pattern Growth: Experiments and Models*. Boston: Kluwer.

3. THE CLASSICAL FAILURE TO ACCOUNT FOR ELECTROMAGNETIC ARROWS OF TIME*

Frank Arntzenius

1. INTRODUCTION

When one uses one's vocal chords, one produces sound waves which spread from one's mouth onto surrounding objects, among which hopefully are some other people's ears. One does not use one's vocal chords so as to dampen sound waves which have spontaneously erupted from surrounding tables, chairs, curtains, walls and ears. However, in standard wave theories the latter process is as much a solution to the equations as the former process, and hence it is at least conceivable that a philosopher might attempt to retract a bad argument in this unusual manner. In this paper, I wish to discuss how classical electromagnetism fails to account for nature's preference for the occurrence of certain types of processes (such as spreading waves) above the time reverse of such types of processes (such as converging waves), and how this is indicative of fundamental problems with the classical picture of the world.

If certain processes occur, but the time reversal of these processes do not, then such processes can be said to establish (or exhibit) an "arrow of time." An arrow of time that is familiar to us concerns the development of the temperatures of pieces of matter. For instance, if a hot brick and a cold brick are brought into contact, and if they are otherwise (thermally) isolated, then they will eventually both become lukewarm. However, initially lukewarm bricks that are brought into contact will not spontaneously become one hot brick and one cold brick.

In the second half of the nineteenth century, L. Boltzmann attempted to reconcile such time-irreversible phenomena with the time reversibility of the laws which he took to govern the behavior of the (molecular) constituents of the bricks. He argued, under the constant fire of opponents (see, e.g., the original papers reprinted and translated in Brush 1965), that indeed the time reverse of *any* process is possible, but that, for example, a spontaneous evolution of a temperature difference between the bricks is extremely *unlikely*, while the evening out of the temperature differences is extremely *likely*. Roughly speaking, Boltzmann justified this claim about the likelihoods of the developments by the assumption that the likelihood of a particular development from particular initial temperatures of the bricks is proportional to the number of initial conditions of the microscopic constituents (molecules) which correspond to that development from those initial temperatures. Thus, on Boltzmann's view the "temperature of bricks' arrow of time" has a statistical origin: The time reverse of every temperature development is possible, but temperature equalization is overwhelmingly more likely than temperature divergence.

Here, I do not wish to discuss in detail the tenability of Boltzmann's position regarding the statistical origin of the arrow of time of classical particle systems. Instead, I wish to discuss whether, provided that one accepts a statistical arrow of time for classical particle systems, one can similarly accept a statistical arrow of time for field-particle systems which satisfy the equations of classical electromagnetism.

For a simpleminded philosopher like me, it would seem most satisfactory if a unified account could be given of all arrows of time. For if not, then one faces additional unexplained facts. It is not too hard to explain why bricks even out their temperatures in the same direction of time as the direction in which metal bars even out their temperatures, if one accepts the same Boltzmannian account of the evening out of the temperatures in both cases. However, if one does not accept the same origin for the spreading of waves as one accepts for the evening out of temperatures of bricks, then one faces an additional problem. Why should waves spread in the same direction of time that temperatures of bricks even out in? Is it a mere fluke? Could waves converge in the direction of time that temperatures even out? In the next five sections, I examine the way in which classical electromagnetism fails to provide an account of certain electromagnetic arrows of time. In the final section, I briefly indicate some reasons why I have the hope of a unified statistical account within quantum field theory of all arrows of time.

In electromagnetism, there is not merely an arrow of time of spreading waves *versus* converging waves. Some fundamental difficulties for classical electromagnetism are closely related to other electromagnetic phenomena which exhibit an arrow of time. That is why we now turn to blackbody radiation.

2. PLANCK, RITZ AND EINSTEIN ON CAVITY RADIATION

Two material bodies can not only come to thermal equilibrium by direct contact, but also by the exchange of radiation through empty space. Equilibria for interacting matter radiation systems exist. The specific problem which Planck was tackling at the end of the nineteenth century was to find the energy density u of radiation in a cavity in matter at temperature T as a function of the frequency f of the radiation. Since cavity radiation can be shown to have the same distribution of energy as the radiation emitted by a perfect blackbody of temperature T, this problem was also known as the blackbody problem.

Not only did Planck want to find $u(f,T)$, he also wanted to show that *any* initial distribution of radiation in the cavity would evolve irreversibly towards such an equilibrium distribution. Planck's derivation of his famous distribution law can be split into two parts (see Planck 1900a, 69-122, for a summary of his previous work; see Planck [1900] 1967a; 79-81, for his first derivation of the Planck distribution law; and see Planck [1900] 1967b, 82-90, for his first "quantum derivation" of the Planck distribution law).

In the first part, Planck considers a number of dipole oscillators placed in a cavity filled with some radiation. He then shows that, given certain assumptions,[1] the result of the interaction of the oscillators and the radiation will be a development towards an equilibrium state for which the average energy $U(f)$ of an oscillator of resonance frequency f will be proportional to the energy density $u(f)$ of the radiation at frequency f: $u(f)=cf^2U(f)$, where c is a constant. In this part, nothing is proved about the distribution of energy over the different frequencies at some temperature T. Indeed, the equilibrium distribution of energy between radiation and oscillators is achieved at each frequency separately without exchanges of energy between different frequencies.

The second part of the derivation is the part where classical physics runs into trouble. According to classical kinetic theory (the equipartition theorem for material oscillators), each oscillator would have an average energy $U=kT$, where k is Boltzmann's constant, when it is in thermal equilibrium with a body at temperature T. In conjunction with the above, this implies the so-called Rayleigh-Jeans distribution law for the radiation: $u(f,T)=cf^2kT$. Unfortunately, this distribution is not only in disagreement with experimental results, it is also absurd. The energy density increases without bound as the frequency gets higher. The total energy in the radiation is infinite. Moreover, given any particular frequency f_0, infinitely more energy is stored at frequencies higher than f_0 than at frequencies below it. Classical electromagnetism plus classical kinetic theory predict a development towards an equilibrium that is nonsensical.

Planck did not assume $U=kT$, nor did he point out the consequences of such an assumption. He posited (several different) expressions for the entropy of an oscillator (in terms of its energy U). From this, he derived (several) distribution laws by maximizing the total entropy under exchanges of energies between the different frequencies. (He did not provide a mechanism for exchanges of energy between different frequencies.) In Planck ([1900] 1967a), he derived the famous Planck distribution law from such an expression for the entropy of an oscillator. This distribution law fitted the experimental data well. In Planck ([1900] 1967b), he justified the expression for the entropy of an oscillator which he had simply posited in Planck ([1900] 1967a). He did this by assuming that the energy of an oscillator of resonance frequency f is composed of an integral number of units of energy $E=hf$, where h is Planck's constant (this is the quantum assumption). He then assumes that the probability of a particular energy distribution over frequencies is proportional to the number of ways in which one can divide the total energy into such (indistinguishable) units and distribute them over the oscillators of different frequencies in accordance with that particular energy distribution. The equilibrium distribution is assumed to be the most probable distribution, and is the so-called Planck distribution.

The irreversible approach to cavity equilibrium is, on Planck's quantum account, nothing but the evolution of a less probable distribution to the most probable distribution. Fluctuations away from equilibrium are possible but unlikely (the larger the unlikelier), and likely to be followed by a return to equilibrium.

In the final section, I suggest that the current version of quantum theory may well provide a sensible statistical account of all electromagnetic arrows of time. However, until then, I wish to concentrate on the troubles that classical electromagnetism has in dealing with electromagnetic arrows of time.

So, let me introduce E. Ritz, who did not accept the quantum solution to the blackbody problem, and in 1908 attempted to come to the defense of the classical account of blackbody radiation (see Ritz 1908). Ritz did not believe that the absurd Rayleigh-Jeans law followed from classical physics, and in particular criticized derivations of Jeans and Lorentz of the Rayleigh-Jeans distribution law from classical electromagnetism and classical statistical physics. In their derivations, Jeans and Lorentz assumed that the radiation field could be *any* solution to Maxwell's equations. Ritz objected to this. He claims that Maxwell's equations allow for both "advanced" (adv) solutions and "retarded" (ret) solutions, and for superpositions of these two types of solutions, but that, in fact, only "retarded" solutions occur in nature.

Informally, retarded solutions correspond to waves that spread from a moving source, and advanced solutions collapse onto a source in coordination with its motion. Formally, consider, for example, the equation $(1/c^2 \cdot d^2/dt^2 - \nabla^2) \, \phi \, (r,t) = 4\pi\rho(r,t)$, where c is the velocity of light, $\rho \, (r,t)$ is the source density, and $\phi \, (r,t)$ is the field amplitude. Given $\rho(r,t)$ as a function of time, $\phi \text{ret} \, (r,t) = \int \frac{\rho(r',t-|r'-r|/c)}{|r'-r|} \, dr'$ is the retarded solution to the above equation. In this case, the field amplitude ϕret at r,t is determined by the source density ρ at position r' at the *earlier* times $t' = t - |r'-r|/c$. Similarly, $\phi \text{adv}(p,r,t) = \int \frac{\rho(r',t-|r'-r|/c)}{|r'-r|} \, dr'$ is known as the advanced solution. In this case $\phi \, adv(r,t)$ is determined by the source density ρ at positions r' at the *later* times $t' = t - |r'-r|/c$. Any superposition $k_1\phi\text{ret} + k_2\phi\text{adv}$ is also a solution to the equation.[2]

Ritz claims that because only retarded radiation in fact ever occurs, Lorentz and Jeans's proof of the Rayleigh-Jeans law is based on a mistaken premise. They mistakenly attribute infinitely many degrees of freedom to the radiation field, and this mistake, in conjunction with the equipartition law applied to the *degrees of freedom of the electromagnetic field* leads to the absurd Rayleigh-Jeans law. Ritz instead advocates a classical retarded action at a distance law in which the field has no (independent) degrees of freedom at all. He does not provide any putative derivation of Planck's distribution law, or some other acceptable distribution law, from such a retarded action at a distance electromagnetic theory. It is indeed not at all obvious how one could do this.

More generally, Ritz claims that the strict time irreversibility of a purely retarded classical electromagnetic theory lies at the very roots of the time irreversibility of the second law of thermodynamics. There is, so he says, nothing statistical about the approach to cavity equilibrium, nor about other arrows of time (see also Ritz and Einstein 1909).

Einstein vehemently disagreed, in print, with Ritz on this point (see Einstein 1909). In the first place, he points out, one would not have conservation of energy for a retarded action at a distance theory since the energy which an accelerated charge radiates into the electromagnetic field is not accounted for in a retarded action at a distance theory. Conservation of energy, Einstein believes, is not a principle to be abandoned unless one has very good reasons (*gewichtige Grunde*) for doing so.

Indeed, Einstein claims, in retarded action at a distance theories the instantaneous state of a system cannot be described at all without using previous states of the system. For example, if a source has emitted a light pulse towards a screen, which has not arrived at the screen yet, then, in the retarded action at a distance theory, this is only represented by the previous states in which the emission took place but by nothing in the current state.

Einstein also points out that the Rayleigh-Jeans law can be derived without an application of the equipartition theorem to the degrees of freedom *of the electromagnetic field* in a cavity. Einstein considers an ion which can oscillate along some axis with resonance frequency f, and which is surrounded by radiation of energy density $u(f,T)$. From Maxwell's theory (Planck's previous work!) one can derive that the average (vibrational) energy $U(f)$ of the ion will satisfy $cf^2U(f)=u(f)$. If this ion is in thermal contact, say, with a gas of temperature T, then it follows from the well-established kinetic theory of heat (for *matter*, e.g., gasses) that $U(f,T)=kT$, and therefore that $u(f,T)=cf^2kT$. Thus even without applying the law of equipartition to *radiation*, the Rayleigh-Jeans law can be derived from classical electromagnetism and the relatively uncontroversial $U=kT$ for a material oscillator in equilibrium with a material body at temperature T.

Since the Rayleigh-Jeans law is absurd, some of the assumptions made in the derivation will have to go. It is simple to see, so Einstein says, how one can modify the above derivation so that the Planck distribution law follows. One need merely modify the statistical theory of heat by the assumption that an ion of resonance frequency f can only have (vibratory) energy which is an integer multiple of hf. Einstein takes the view that the irreversible development towards cavity equilibrium consists of a transition from a less probable to a more probable (macroscopic) state. Planck's definition of the probability of a state in terms of the number of complexions corresponding to such a state can then be used to derive such an approach to a sensible, and experimentally satisfactory, equilibrium state. Einstein remarks that, conversely, the experimental evidence can be viewed as an empirical indicator of the probability of states.

The blackbody arrow of time, on Einstein's view, is statistical in origin, and this must be so because of the reversibility of all the fundamental laws of physics. Classical physics does not provide a sensible matter-radiation equilibrium, let alone a statistical or nonstatistical approach to matter-radiation equilibrium. Ritz's attempt to rescue a classical account of the blackbody arrow of time yields no positive results. Moreover, it is based on an attack on an assumption which, as Einstein points out, in fact is not needed in the derivation of the dreaded Rayleigh-Jeans law from classical electromagnetism.

Now let us turn to another electromagnetic arrow of time which has provided classical electromagnetism with a headache.

3. RADIATION DAMPING

Electromagnetic waves carry energy (and momentum) in classical electromagnetism. Hence, an accelerated charged particle should be losing some energy to the field as it emits electromagnetic waves. In fact the particle should pay for this loss of energy in terms of its kinetic energy. Thus the acceleration of the charged particle due to some applied external force should be somewhat less than it would be if there were no electromagnetic radiation emitted by the charged particle. If one would apply the same force to two particles of identical mass, the one charged and the other neutral (having neutral elementary constituents), then the charged particle should accelerate less than the neutral particle. This damping of the accelerated motion of a charged particle is known as radiation damping. If one were to retain Newton's law $F_{total} = m.a$, then the damping of the acceleration must be associated with a damping force. One cannot have differing accelerations of particles with identical mass, if the total forces on the particles are identical, without violating Newton's second law.

The origin of this radiation damping force was a big puzzle. An answer to this puzzle was first proposed by H. Lorentz. He proposed that charged particles not only feel the effect of the electromagnetic fields of other particles, but that they also feel the effect of their own field. For an extended charged particle such a view is indeed natural. If some external force is applied to one part of, say, an electron, then the charge in that part of the electron will produce an electromagnetic wave which, due to the finiteness of the speed of electromagnetic waves, will reach the other parts of the electron later. One can show that the total effect of such fields on the electron in the presence of an external applied force is nonzero. However, note that such a calculation is made on the *assumption* that the waves travelling in the interior of the electron are the retarded waves and not the advanced waves. Thus Lorentz's account and other such accounts can only be construed as an attempt to locate the origin of the damping force, not as explanation of its retarded (damping as opposed to antidamping) nature.[3]

The result of Lorentz's calculations is that the total force which an electron exerts on itself equals the sum of three terms. The first term is proportional to the acceleration of the electron. When this term is added to Newton's second law one obtains $F_{ext}=(2e^2/3rc^2+m)a$. This first term is known as the electromagnetic mass term since it behaves like a mass term in Newton's second law. Lorentz regarded this term as a great advantage since he could claim that *all* of the inertial behavior of matter was due to the electromagnetic mass of charged particles. He indeed suggested that one remove the mechanical mass m altogether from Newton's second law. Given the charge e of an electron, the radius r of an electron and the speed c of electromagnetic waves, one could calculate the electromagnetic mass of the electron. Conversely, given the electromagnetic mass, e and c, one can calculate the radius r of the electron. This figure is known as the classical size of the electron and is approximately 3.10^{-13} centimeter. The second term in Lorentz's result is the sought after damping term. It is exactly the same term as one obtains by calculating the force which is needed to slow the electron down in accordance with the energy it emits and the conservation of total energy (at least for periodic motions of the electron). The equation of motion for a charged particle to which an external force is applied becomes $F_{ext}+F_{damp}=F_{ext}+2e^2/3c.^3da/dt=(m_{em}+m_{mech}).a$, where the mechanical mass m_{mech} can be removed if one agrees with Lorentz that the electromagnetic mass m_{em} can account for the inertial behavior of charged particles. This damping law is a time-irreversible law: If one reverses the velocity of a charged particle and one reverses the time development of the applied external force, then its motion will *not* be reverse of the original motion. The assumption of self-interaction yields exactly the damping force one was seeking on the basis of total energy conservation and Newton's second law. So far then we have a triumph. Let us turn to the problems.

(i) The third term in Lorentz's calculation corresponds to a sum of higher order terms which are proportional to fourth and higher time derivatives of the position of the electron. For violent changes in the motion of an electron such terms can become significant. They cannot be eliminated for a finite sized electron.

(ii) The Lorentz-electron is highly unstable. The electrostatic (Coulomb) forces of parts of the electron on other parts of the electron would explode the electron if they were the only forces present. H. Poincaré suggested a remedy for this by introducing a binding force which would keep the electron stable. He did this by postulating so-called Poincaré-stresses in the form of a particular stress-energy tensor.

(iii) The equations of motion which result from Lorentz's calculations and Newton's second law, even ignoring the higher order terms, have so-called runaway solutions. For instance, when one applies no external force to a single charged particle the particle can remain at rest, or it can accelerate away with always exponentially increasing acceleration! Such runaway solutions can be avoided by imposing an initial condition which ascribes a certain value to the initial acceleration of the charged particle. However, the magnitude of the initial acceleration $a(0)$ will be a functional of the total future development of the externally applied force F_{ext}: $a(0)=1/vm \int_0^\infty \exp(-t/v).F_{ext}(t)dt$, where $v=2e^2/3mc^3$. Thus, for instance, if one "hits" a particle with a delta-force at a time $t=0$ then, so as to avoid a runaway behavior, the particle must begin to accelerate before $t=0$. In fact it must begin to accelerate infinitely long before any external force is applied; it preaccelerates. The fact that one is even free to specify an initial acceleration is another strange property of the equations of motion reached at by Lorentz. Normally one is only free to specify the initial position and velocity of particles.

(iv) Atoms are unstable. As Bohr (1913) pointed out in his first work on the (old) quantum theory of spectral lines, one electron (or a few electrons) in an orbit around a nucleus would very quickly radiate away its energy and collapse onto the nucleus if radiation damping were assumed. This problem was known before Bohr's remarks. However, in previous atomic models, for example, those of Thomson, very many electrons orbited around even the lighter nuclei (or in the positive charge cloud). An orbiting large group of electrons radiates away far less energy per unit time than a single electron or a few electrons. Before the scattering experiments of the early twentieth century (and Thomson's and Rutherford's analyses of the results) indicated that the lighter elements contained relatively few electrons, the problem of collapsing electrons was not so immediate. Of course, it is still a problem in principle for any model that does not allow the electrons and the positive charge to be in *static* equilibrium. It is especially so because classical electromagnetism *rules out* any *static* particle equilibrium (by Earnshaw's theorem—see, e.g., Pais 1986; 181)!

One might hope to solve problem (iv) by pointing out that incoming radiation (emitted by neighboring atoms) could be (partially) absorbed by an orbiting electron and thus increase its energy. Indeed one might hope that some equilibrium with some average orbit might result from absorption and emission. However, this is exactly the blackbody problem. Classical electromagnetism predicts that, no matter what the precise details are of one's atomic model, all the energy will eventually disappear into (the high frequency range of) the electromagnetic field, and the electrons will be bound to collapse after all.

In classical electromagnetism, all charged particles should eventually collapse onto oppositely charged particles due to radiation damping while radiating away huge amounts of energy. This constitutes an arrow of time that is not only in disagreement with some refined experiments, but indeed contradicts some of the more obvious features of our universe.

(v) Between 1905 and 1910 Einstein's work, and certain experimental results to which he referred, made it plausible that in fact the energy of radiation emitted by an atom during an elementary emission process does not disperse over larger and larger areas in space as time elapses after the emission. He argued that the energy remains so concentrated (directed) so as to be completely regainable during an elementary absorption process by one other atom.

Some of the above objections to classical electromagnetism with radiation damping can only be removed by the assumption that the electron is not extended in space, but that it is a point particle:

(i) The higher order terms disappear (go to zero).

(ii) One may just assume that the electron is stable, one does not need to invoke Poincaré-stresses.

(iii) Runaway solutions remain. Preacceleration remains.

(iv) Electrons will collapse onto nuclei.

(v) Energy will disperse.

To this list one has to add:

(vi) The electromagnetic mass of the electron becomes infinite. One can remedy this by ignoring (subtracting away) Lorentz's first term, and reintroducing good old mechanical mass. Such a procedure is known as classical mass renormalization.

Thus we see that the classical electromagnetic account of radiation damping is in deep trouble; not only is it in conflict with certain subtle experimental results, but it is in deep theoretical trouble: It yields results which, if not inconsistent, cannot describe a universe even remotely like our own. Let us also reiterate that Lorentz's account of radiation damping in no way explains the origin of the electromagnetic arrow of time. Lorentz's calculation yields a direction of time as can be seen from time-irreversibility of the law of radiation damping, but he derives this *assuming* the retarded nature of the electromagnetic waves which he uses to calculate the total force of an electron on itself.

Let us now turn to attempts to account for the presumed preference of nature for retarded radiation within classical physics.

4. THE ORIGIN OF THE RETARDED NATURE OF RADIATION

K. Popper has expressed the opinion that an account of the retarded nature of radiation can be given in classical physics. His account, he claims, is not statistical. According to him, the reason why waves spread is not the same reason as the reason why temperatures of material bodies even out:

> Suppose a film is taken of a large surface of water initally at rest into which a stone is dropped. The reversed film will show contracting circular waves of increasing amplitude. Moreover, immediately behind the highest wave crest a circular region of undisturbed water will close in towards the centre. This cannot be regarded as a possible classical process. (It would demand a vast number of distant coherent generators of waves the coordination of which, to be explicable, would have to be shown, in the film, as originating from one centre. This, however, raises precisely the same difficulty again, if we try to reverse the amended film.) Thus irreversible processes exist. (On the other hand, in statistical mechanics all processes are in principle, reversible, even if the reversion is highly improbable.) (Popper 1956, 538)

To claim that the coordination of the generators of the waves would require organization (Popper means *previous* organization) from a center is to prejudge the issue. If Popper is happy to explain the coordination of amplitudes on a circle in the case of the outgoing wave in terms of the previous dropping of the stone, then he should also be happy to explain the coordination of the amplitudes in the case of the converging waves in terms of the later popping out of water of the stone. Not accepting such an explanation is to presuppose a direction of time over and above the presented phenomena. But the point was to demonstrate an expected time asymmetry in the phenomena. Popper is prejudging the issue by first requiring an "explanation" for the initial state under consideration, and then implicitly assuming that such an "explanation" can only refer to an *earlier* state which he somehow considers less problematic than the state under consideration. In any case, to talk of good and bad "explanations" for a particular set of initial conditions seems a dubious enterprise to me. Any previous or later state which, given the laws of physics, implies the state to be "explained" seems good enough, unless one wants to prejudge the issue of the time-asymmetry.

A more explicit account of wave retardation within classical physics, similar to Popper's account, has been put forward by O. Penrose and I. Percival (1962). They impose, in effect, the requirement that correlations between states in space-like separated spacetime regions do not occur unless there is a common cause of these correlations in the *past* lightcone of the regions. In particular, they consider a dipole oscillator which oscillates between times t_0 and t_1, but does not oscillate outside these times. They then show that, if states in space-like separated regions can only be correlated through a *past* common cause, then the oscillation must be accompanied by retarded radiation.

Before criticizing their account, let me note that they call their account a *statistical* account. They do so because their account is phrased in terms of the ruling out of *correlations* between states in different regions. Nevertheless, their account is not statistical, in the sense that advanced radiation on their account is ruled out completely rather than attributed some very low probability.

As we have seen in Section 3, associated with a strict retarded radiation law, is a strict law of radiation damping. If one wishes to avoid runaway solutions then one finds oneself reintroducing a new form of "inexplicable coordination": the coordination of the force to be applied at a later time and the "preacceleration." The external force could be caused by a field disturbance which reaches the dipole at a particular time and then exerts a force on the particle. There would then be an inexplicable, uncaused correlation between the field disturbance propagating to the dipole and the preacceleration of the dipole.

The reason why this coordination does not show up in Penrose and Percival's account (nor indeed in Popper's account) is that they do not include an account of the force needed to set the dipole in its motion. The dipole is simply *assumed* to oscillate for a period, but the external force, and the source of this force, which is needed to oscillate the dipole, is not considered in Penrose and Percival's paper. If they would include it, and if they would ban runaway solutions, then they would find a correlation between the preacceleration of the dipole and the incoming field which acts on the dipole. Indeed they would find that this correlation has no common cause in the past lightcones.

Until Popper, Penrose and Percival include in their accounts the source of the force which is needed to set into motion the source of the retarded radiation; they have not shown that one can rule out advanced radiation by the ruling out of correlations which do not have a common cause in the past.

Let us now turn to an attempt to provide a statistical account of the damping arrow of time within classical physics which does not assume retarded solutions only.

5. TIME-SYMMETRIC CLASSICAL ELECTROMAGNETISM

Feynman and Wheeler (1945) assume that the radiation associated with the motion of a charged particle consists of half advanced and half retarded radiation. This of course is consistent with the fundamental equations of classical electromagnetism, but it is a stronger theory than just those equations since it rules out many solutions to the fundamental equations (such as pure retarded or pure advanced solutions). They furthermore assumed that charged particles are point particles which do not interact with their own fields. To calculate the force on a charged particle one need only evaluate the fields of other particles at the position of the particle under consideration.

To get a model for the phenomenon of radiation damping they consider a single charged particle, the source, surrounded by a completely absorbing medium, that is, a collection of absorbing particles such that there never is any electromagnetic radiation outside the absorber. They then show that the strength of the field acting on the source will give rise to exactly the damping force which one desires (Lorentz's second term), at least if one makes one additional assumption, which they claim to be a statistical assumption. This assumption is that the retarded fields coming from the absorber (due to the motions of the particles in the absorber) sum to zero at the position of the source, at the moment of its acceleration by an applied external force. If instead of this one would assume that the sum of the advanced waves "originating" from the absorber (at a later time) sum to zero at the position of the source at the moment that it is accelerated by the external force, then one would arrive at an antidamping force given by the time reverse of the law of radiation damping. To obtain the desired damping law one has to assume that the retarded wave emitted by the absorber particles, which "travels" in the forwards direction of time, and arrives at the moment of acceleration, yields a net zero force, whereas one must assume that the advanced wave emitted by the absorber particles which "travels" in the backwards direction of time, and arrives at the source at the moment of acceleration, does not sum to zero. Feynman and Wheeler claim that this follows from statistical considerations:

> Evidently the explanation of the one-sidedness of radiation is not purely a matter of electrodynamics. We have to conclude with Einstein that the irreversibility of the emission process is a phenomenon of statistical mechanics connected with the asymmetry of initial conditions with respect to time. In our example the particles of the absorber were either at rest or in random motion before the time at which the impulse was given to the source. It follows that in the equation of motion the sum $\Sigma\,F_{ret}$ of the retarded fields of the absorber particles had no particular effect on the source. Consequently the normal term of radiation damping dominates the picture. In the reverse formulation of the equation of motion, the sum of the advanced fields of the absorber particles is not at all negligible, for they are put into motion by the source at just the right time to contribute to the sum $\Sigma\,F_{adv}$. This contribution, apart from the natural random effects of the changes of the absorber, has twice the magnitude of the usual damping term. The negative reactive force is therefore cancelled out, and a force of the expected sign and magnitude remains. (1945, 170)

Feynman and Wheeler's idea is that statistical mechanics justifies the assumption that the initial motion of the absorber particles is chaotic so that the retarded waves resulting from this chaotic motion can be expected to have a zero net result at the position of the source. However, they argue that the advanced radiation coming from the absorber particles will not sum to zero since the absorber particles will have been set into nonrandom motion by the retarded wave coming from the source (advanced radiation "leaves" the absorber particles *after* the source has been set into motion and arrives at the source at exactly the time that it is set into motion). However it would seem that such an argument within time-symmetric electromagnetism is at the very best question-begging. For if the retarded waves coming from the source can set into nonrandom motion the absorber particles at a later time, then the advanced waves coming from the source should similarly be capable of producing nonrandom motion in the absorber particles at the earlier time. Indeed if one can reason that the retarded field of the source will result in nonrandom motion of the absorber particles at the later time which in turn results in advanced radiation, which, travelling backwards in time to the source, yields a damping force on the source (at the time of the acceleration), then similarly one can reason that the advanced field of the source will result in nonrandom motion of the absorber particles at an earlier time, which in turn, through a retarded wave, will yield an antidamping force on the source at the time of its acceleration. But the two conclusions are inconsistent, and hence the arguments must be rejected.

At the very best one can just prescribe initial random motion in the absorber and later nonrandom motion in the absorber, foregoing any "explanation" of this fact. But this is hardly an account of the arrow of time of electromagnetism in terms of statistical mechanics. In fact I am more inclined to worry about a transition from earlier randomness to later nonrandomness of the motion of the particles *in view of statistical mechanics*, than regard it as *warranted by statistical mechanics*. Another worry concerns the consequences of the later nonrandomness at even later times, or the earlier randomness at even earlier times. For instance, let us assume that the absorber particles are all at the same distance from the source, that they are on the surface of a sphere centered on the source. A nonrandom motion which leads to a nonzero advanced sum at an earlier time on the source (yielding a damping force) will similarly lead to a nonzero retarded sum at a later time if the source is in the same position, yielding an antidamping force, which is not what one wants.

Regarding the problems which I indicated above for classical electromagnetism which assumes retarded radiation only, Feynman and Wheeler are rather successful:

(i)   There are no higher order terms in their theory.

(ii) They may assume their point particles to be stable.

(iii) They have runaway solutions in the case of complete absorption. They have preacceleration, but claim that advanced effects are to be expected in a time-symmetric theory.

(iv) Electrons will collapse onto atoms.

(v)   Energy will disperse.

(vi) Self-energy is zero (there is no self-interaction).

Regarding points (iv) and (v) it should also be added that Feynman and Wheeler always intended to quantize their theory, so that it is no objection against *their* enterprise that it leaves unsolved problems which they only expected to solve in the quantized theory. However we will see in the next section that the quantum account of the electromagnetic arrow of time is different from the classical account so that I include these problems as problems faced by a classical time-symmetric electromagnetic theory.

6. SUMMARY OF CLASSICAL PROBLEMS WITH
ELECTROMAGNETIC ARROWS OF TIME

A) If one only allows purely retarded solutions, then pure radiation damping has to be assumed, with the associated problems labeled (i) through (vi) in the previous section. These problems range from disagreement with subtle experiments—for example, photoelectric effect w.r.t. (v)—to fundamental theoretical and conceptual problems—for example, instable matter and preacceleration.

B) The *very* intuition which one might attempt to use to justify the retarded nature of radiation is violated by preacceleration, which itself is a consequence of the assumption of retarded radiation (and the "no runaway" assumption).

C) I am not aware of any account, other than the Feynman-Wheeler account, which claims that radiation damping is merely more *likely* than antidamping in classical electromagnetism. If one does not accept Feynman and Wheeler's account one has to wonder why waves spread in the same direction in which temperatures of material bodies even out. (One could argue whether Penrose and Percival's account, were one to accept it, offers a unified account of the particle and field arrows of time.)

D) Feynman and Wheeler allow waves "travelling" in both directions of time. But, since they do, a common cause of a coordinated motion in the absorber could both exist before and after that coordinated motion takes place. Hence there is no reason to assume that the randomness, which they need to establish a statistical account of radiation damping, exists in the past but not in the future.

E) When forced to use the statistics of the electromagnetic field in order to solve the blackbody problem, classical electromagnetism yields an absurd answer.

One could conjecture that at least part of the problems due precisely to the fact that it is so difficult, if not impossible, to give a statistical account of the arrow of time of electromagnetic phenomena within classical electromagnetic theories. Classical physics fails because it cannot handle the real issue: to give an account of electromagnetic arrows of time in terms of the statistics of the initial states of the particles *and the initial field configurations*. Let us now conclude by briefly considering a theory which faces the problems head-on by treating electromagnetic radiation in such a way that its statistics are on a par with the statistics of particles.

7. PROSPECTS FOR ELECTROMAGNETIC ARROWS OF TIME
IN QUANTUM ELECTRODYNAMICS

In this section I roughly indicate why I believe that quantum field theory overcomes the classical failure by providing a unified statistical account of electromagnetic arrows of time.

7.1. *Cavity Radiation.*

In quantum electrodynamics the energy levels of both "field" systems and "particle" systems are quantized. In quantum statistical mechanics, one (usually) makes the fundamental assumption that the probability that a system, which is in thermal equilibrium with a "reservoir" at temperature T, is in energy level Ei, is proportional to $\exp(-Ei/kT)$, where k is Boltzmann's constant. Given the quantization of the electromagnetic field, one can easily derive from this assumption Planck's cavity distribution law (see for instance Feynman and Hibbs 1965, chap. 10; or Loudon 1983, chap. 1). (This quantum mechanical method corresponds to the classical "Jeans-Lorentz" derivation: One works directly in terms of the field.) Alternatively, one can derive Planck's distribution law from the fundamental assumption, the quantization of the energy levels of material oscillators, and the emission and absorption probabilities of photons by material oscillators (see, e.g., Harris 1971, chap. 2). (This corresponds to the "Planck-Einstein" derivation: First establish the field-material oscillator equilibrium, then work with the material oscillators.)

The fundamental assumption applies both to "particles" and to "fields" in quantum field theory. The justification for the fundamental probabilistic assumption, and indeed for the assumption that a system originally not in thermal equilibrium with a reservoir, will most likely move towards equilibrium, may be sought in quantum ergodic theorems and/or quantum H-theorems (see for instance Jancel 1969). However, that is a problem which concerns the foundations of Boltzmann's statistical approach, and that problem is not the subject of this paper. Whatever problems there are with Boltzmann's approach, the point is that the cavity radiation arrow of time can be accounted for in quantum statistical mechanics in the same way as the pure particle arrow of time. On this account the cavity arrow of time consists of an extremely likely approach to a sensible and experimentally satisfactory equilibrium.

7.2. *Radiation Damping.*

Radiation damping is the phenomenon that charged particles which are accelerated by some external force emit electromagnetic radiation and pay for this in terms of their energy (and momentum). In quantum electrodynamics this phenomenon shows up in several ways. A free charged particle which exchanges a virtual photon with some other source, that is, a charged particle subjected to an external electromagnetic force, will have a certain probability amplitude for the emission of a real photon. By energy conservation, it will have to pay for the emission of such a real photon in terms of its energy. Indeed by applying energy and momentum conservation one can show that a charged particle not exchanging a virtual photon cannot emit a real photon: A charged particle not subject to an external force cannot emit (real) electromagnetic radiation. One might wish to call the emission of a real photon, when exchanging a virtual photon with an external source, and

the accompanying change in energy and momentum of the charged particle, the emission of retarded radiation accompanied by radiation damping.

On the other hand, due to the time reversibility of quantum electrodynamics, a charged particle which exchanges a virtual photon with an
external source also has a certain probability amplitude to absorb a real
photon, and in the process conserve total energy and momentum. One
might wish to call such a process the absorption of advanced radiation
accompanied by "antidamping." Of course the total probability for such an
absorption depends on the number of real photons present (on the probability amplitudes for the numbers of real photons, to be more precise).
Thus one finds a particular probability amplitude for radiation damped
motion and a particular probability amplitude for radiation antidamped
motion, where the probability amplitudes depend on (the probability amplitudes of) the initial particle and field states.

Consider for instance a group of unbound charged particles which can
exchange virtual photons, and emit and absorb real photons. Any initial
state is likely to develop towards an equilibrium state in which, on average, equally many real photons are absorbed as emitted. "Damped" behavior is likely to occur if one is initially in a state with less photons than in
the equilibrium state; "antidamped" behavior is likely to occur if one is
initially in a state with more real photons than in the equilibrium state.

Similarly, consider a group of hydrogen atoms. Whether the electrons in
particular bound states are likely to emit or likely to absorb radiation
depends on the number of real photons flying about (at the relevant
frequencies). In fact, solving the likelihoods of "damped" and "antidamped"
behavior for a group of atoms in a reflecting box is just solving the blackbody problem.

A large group of accelerating electrons, surrounded by relatively few real
photons, will most likely emit photons (radiation) in all directions, and have
their motion damped. This is a small step towards the elusive goal of equilibrium. The earth and its environment are so far from equilibrium that one can
actually broadcast radio and television this way. However, this does not mean
that accelerated charged particles always emit photons. In the early universe television and radio broadcasts did not and could not occur.

Within quantum electrodynamics the occurrence of "damped" behavior,
or the occurrence of "antidamped" behavior, can be explained in terms of
a likely development from a given initial state towards an equilibrium
state.[4]

7.3. *Retarded versus Advanced Radiation.*

If one takes the perturbative approach to quantum field theory seriously,
as I have done,[5] then the distinction between advanced and retarded
radiation evaporates. A real photon as it is emitted might be called retarded
radiation, and that same photon, as it is absorbed by another charged

particle, might be termed advanced radiation. I have tried to accommodate for this by speaking of the emission as retarded *behavior*, and of the absorption as advanced *behavior*, but even that is rather inappropriate terminology. In quantum electrodynamics retarded behavior at some time and place, in some solution to the equations, does not indicate retarded behavior at all times and places, in that solution. If there is some unlikely fluctuation from equilibrium at some time, and it takes advanced behavior to return to equilibrium, then advanced behavior is most likely to follow. If it takes retarded behavior then retarded behavior is likely to follow.

There is a situation in which one could still insist that one could properly speak of a retarded solution, which displays retarded behavior at *all times*, and of an advanced solution, which displays advanced behavior at *all times*. Consider a universe with finitely many particles forever confined to a finite region. In such a case one might call a solution, in which no radiation is travelling towards the group of particles from past infinity, a retarded solution, and a solution in which no radiation travels outwards from the group of particles to future infinity, an advanced solution.

Yet again, which form of radiation is likely to occur depends on the initial state of fields and matter. It depends on the initial energy levels of the particle states and the abundance of photons of different frequencies. If there are initially very few photons and the matter is predominantly in a highly excited state, then, initially, photons are likely to be emitted to future infinity. If, initially, there are many photons and the matter is predominantly in ground states, then, initially, photons from past infinity are likely to be absorbed. This can be stated in a different way: An initially hot finite system of material particles in an initially cold bath of photons will emit radiation until the temperature of the matter becomes equal to the temperature of the photon bath. Similarly, for initially cold matter in a hot photon bath one has a likely evolution towards temperature equilibrium.

If the temperature of the photon bath is initially zero, then one will find a pure retarded solution, and an evolution towards zero temperature of both photons and matter. The reason that the temperature of the photon bath is not raised by the photons which are emitted is that we have assumed a finite amount of matter in an infinite empty space. If one had either placed a finite amount matter in a finite empty space with reflecting walls, or an infinite amount of matter in an infinite space, then the temperature of the photon bath would have been raised by the emitted photons, and the temperature would have evened out at some nonzero equilibrium temperature. But then one would not have pure retarded or pure advanced solutions.

For the same reason, one cannot have a pure advanced solution for a finite amount of matter in an infinite space. The matter would quickly become just as hot as the incoming radiation and an equilibrium between emitted radiation and absorbed radiation would evolve.

Our own universe seems to be in the situation where the temperature of large chunks of matter is much hotter than that of the surrounding radiation. Therefore stars emit "retarded" radiation. This behavior is the likely evolution towards the elusive goal of equilibrium.

The ultimate origin of this nonequilibrium is another matter that is not part of this paper. It may be that self-gravitating systems have no proper equilibrium states. It may be that thermodynamic equilibrium is broken by the expansion of the space in which the equilibrium obtains, and so on. The main claim of this last section does not hinge upon answers to such problems. The main claim is that there is hope that all electromagnetic arrows of time can be accounted for in a uniform statistical manner within quantum field theory, whereas there is very little hope that classical electromagnetism can provide a satisfactory account of electromagnetic arrows of time.

University of Southern California

Notes

*Most of the research for this paper was done while I was receiving financial support from the University of Pittsburgh, for which I am most grateful.

1. Two interesting assumptions are the following:

(1) The oscillators satisfy a "radiation damping" equation of motion, which is time *irreversible*. How one can get a time irreversible "radiation damping" equation of motion from time-reversible fundamental equations will be discussed in Section 4.

(2) Planck assumes a certain randomness condition on a Fourier decomposition of the radiation. This is the assumption of "natural radiation." This assumption guarantees that the behavior of an oscillator in the radiation is determined by certain "course grained" properties of the radiation. Whether this assumption can be construed as a bona fide statistical assumption is an interesting question which will not be addressed in this paper.

2. To be able to write the solutions to the equation in the stated forms, one must assume that the sum of the contributions to the field amplitude at a point x from sources at a distance greater than L from x, goes to zero as L goes to infinity. This is automatically insured if one confines all sources to a finite region in an infinite space. Whether advanced solutions or retarded solutions (or superpositions thereof) occur then depends on whether source free radiation is coming in from past infinity, or whether radiation escapes at future infinity. See also Sciama (1961).

3. Lorentz himself was interested in giving an electromagnetic account of the mass and size of an electron, and of the damping force associated with the radiation emitted by an accelerated electron. He took it for granted that the radiation associated with the motion of the electron reaches other points in space *after* the associated motion and not *before*. He was not in any way attempting to *account* for this time asymmetry. See Lorentz (1952).

4. There are other phenomena that are usually associated with radiation damping in quantum field theory. For instance the line broadening of spectral lines is usually considered to be a consequence of radiation damping. Such phenomena are not relevant for this paper since no arrow of time is associated with them. (The line broadening occurs in emission as well as absorption spectra.)

5. That is to say, I have taken the results of calculations of perturbative quantum electrodynamics for the probability amplitudes of the transition from certain "in-states" to certain "out-states" seriously. Whether one takes the Feynman diagrams seriously, and speaks of exchanged virtual photons, is irrelevant to my story. One need merely accept that to have an external force exerted on a charged particle is to have nonzero probability amplitudes for an "out-state" that is not equal to the "in-state"; whether one actually forms a picture of what happens in the interaction region is irrelevant.

References

Bohr, N. (1913), "On the Constitutions of Atoms and Molecules," *Philosophical Magazine,* vol. 26, pp. 1-25.

Brush, S. (1965), *Kinetic Theory, Selected Papers,* vol. 2. Oxford: Pergamon Press.

Einstein, A. (1909), "Zum gegenwartigen Stand des Strahlungsproblems," *Physikalishe Zeitschrift 10* (No. 6): 185-93.

Feynman, R. and Hibbs, A. R. (1965), *Quantum Mechanics and Path Integrals.* New York: McGraw-Hill.

Feynman, R. and Wheeler, J. (1945), "Interaction with the Absorber as the Mechanism of Radiation," *Reviews of Modern Physics 17*: 157-81.

Harris, E. G. (1971), *A Pedestrian Approach to Quantum Field Theory.* New York: Wiley.

Jancel, R. (1969), *Foundations of Classical and Quantum Statistical Mechanics.* Oxford: Pergamon Press.

Lorentz, H. A. (1952), *The Theory of Electrons.* New York: Dover Publications.

Loudon, R. (1983), *The Quantum Theory of Light.* Oxford: Clarendon Press.

Pais, A. (1986), *Inward Bound.* Oxford: Oxford University Press.

Penrose, O. and Percival, I. (1962), "The Direction of Time," *Proceedings of the Physics Society 79*: 605-16.

Planck, M. (1900), "Uber irreversibile Strahlungsforgange," *Annalen der Physik 1*: 69-122.

————. ([1900] 1967a), "On an Improvement of Wien's Equation for the Spectrum," in D. Ter Haar (ed.), *The Old Quantum Theory.* Oxford: Pergamon Press, pp. 79-81.

————. ([1900] 1967b), "On the Theory of the Energy Distribution Law of the Normal Spectrum," in D. Ter Haar (ed.), *The Old Quantum Theory.* Oxford: Pergamon Press, pp. 82-90.

Popper, K. (1956), "The Arrow of Time," *Nature 177*: 538.

Ritz, W. (1908), "Uber die Grundlagen der Elektrodynamik und die Theorie der schwarzen Strahlung," *Physikalische Zeitschrift 9* (No. 25): 903-7.

Ritz, W. and Einstein, A. (1909), "Zum gegenwartigen Stand des Strahlungsproblems," *Physikalische Zeitschrift 10* (No. 9): 323-24.

Sciama, D. (1961), "Retarded Potentials and the Expansion of the Universe," in T. Gold (ed.), *The Nature of Time*. Ithaca: Cornell University Press, pp. 55-67.

4. THE REORIENTATION OF NEOCLASSICAL CONSUMER THEORY*

Edward J. Green and Keith A. Moss

I. INTRODUCTION

In neoclassical microeconomics, the analysis of consumers' behavior is based on a theory of utility maximization. It is well known that, in the century since the beginning of the "marginalist revolution," the interpretation of this theory has changed markedly. In this paper we attempt to provide an understanding of why, during the early part of this century, such a change occurred. We will also discuss why the specific conception of the theory that dominated microeconomics during the middle third of the century has subsequently receded in influence.

Let us briefly recapitulate the history of events with which we will be concerned. The early marginalists tended to think of utility in much the same terms as had previous philosophical utilitarians: that is, as a quantity that could be compared cardinally at least within an individual, and perhaps across individuals. Progressively this version of utility was replaced by a more parsimonious utility concept, admitting only ordinal, intrapersonal comparisons. A new formal theory, in which an ordinal ranking of alternative consumption bundles is taken to represent the consumer's preferences, was shown to be equivalent (under mild restrictions) to this ordinal utility theory. The expected-utility theory of decision under uncertainty, which had initially been thought to depend essentially on a representation of intensity of preferences, was shown actually to be axiomatizable within the ordinalist framework. The observable implications of ordinal-utility theory regarding an individual's budget-constrained consumption choices were completely characterized within revealed-preference theory, and within the related theory of integrability of demand functions.

This reformulation of the utility concept had two aspects. First, it purged the concept of some of its aspects—particularly, cardinality and interpersonal comparability—that proved to be superfluous to consumer theory. Second, it replaced a theory that was rather informal, and perhaps even somewhat ambiguous, with an explicit and tightly circumscribed one. Historically, the reformulation was essentially complete by the early 1950s. There were two alternative formulations of the theory: ordinal utility theory which had reached its culmination in the work of Pareto ([1909] 1971), Slutsky ([1915] 1952), and Hicks and Allen (1934), and Samuelson's revealed-preference theory augmented by the Strong Axiom of Revealed

Preference formulated by Houthakker (1950).[1] We would characterize the Hicks-Samuelson program as having two salient features. First is a conception of the formal structure of the theory as a single, comprehensive, axiomatizable theory. Second is a pair of heuristic criteria for the adoption of assumptions regarding consumers' preferences. The first criterion is that axioms designed to assure in a simple, parsimonious way the *mere consistency* of consumers' choice behavior may be adopted without systematic consideration of their evidential support. The second criterion is that assumptions designed to place *substantive restrictions* on the patterns of consumers' choices may be adopted if, in principle, their truth could be decided by observing the purchases of an individual household as its wealth, and the prices that it faced, varied.

Several recent philosophical studies of the Hicks-Samuelson theory (e.g., Rosenberg and Wong) have assessed this research program in consumer theory in negative terms, and even sympathetic philosophers (e.g., Hausman 1981) have characterized it as being successful only in a limited sense, and with respect to a circumscribed set of goals. In this paper, we propose to reconsider those conclusions. We will concur that this research program was, in some sense, a scientific failure. However, we will emphasize the gravity of the issues that led these economists to adopt the sort of theory and heuristics that they did adopt, and the large degree of success that they achieved in addressing those issues. This emphasis differs sharply from a view that the Hicks-Samuelson program did not conform to any coherent model of scientific practice, or that it was the outcome of adherence to a model (such as naive verificationism) that is coherent but simplistic. Rather, on our view the two salient features of the theory were reasonable solutions to the problem of formulating a conceptually satisfactory theory of consumers' decisions, but this way of assuring conceptual adequacy left the theory too weak to address successfully the particular scientific problems that it would have had to handle in order to be considered successful. This is not a situation that the creators of the theory could have forecast and avoided.

II. THE FORMULATION OF A THEORY: JEVONS

The research with which we are most directly concerned was undertaken approximately half a century after neoclassical consumer theory was formulated by Edgeworth, Jevons, Walras, and others. This research constituted an attempt to provide more satisfactory foundations for that theory. For this reason, we begin our study by describing the received version of the theory, and by examining in what respect its foundations were suspect. It is convenient to do so by discussing the work of two prominent contributors to the theory, Jevons and Pareto. This discussion falls far short of being a complete history. Its modest goal is to provide a sense of how one of the founders of the theory conceived of what he was doing, and of the foundational problems engendered by the subsequent logical reconstruction of that conception.

The first edition of William Stanley Jevons's *The Theory of Political Economy* was published in 1871, and the second in 1879. This was roughly sixty years after classical economic theory had attained its mature form. (David Ricardo's *Principles of Political Economy and Taxation* had been published in 1817. John Stuart Mill's *The Principles of Political Economy*, published in 1848, was in most respects an elaboration of Ricardo's theory. A notable exception is that Mill endorsed Malthus's theory of population, against which Ricardo had argued vigorously.) Two components of the classical theory that are particularly relevant to Jevons's theory are the labor theory of value and the theory of rent. The labor theory of value consists of two principal assertions. The first is that the relative price of two goods is the ratio of the marginal costs of their production. (Linear production functions were typically assumed for manufacturing, so there was no need to distinguish among the various cost concepts. In the theory of rents, where marginal cost and average cost are distinct, it is clear that marginal cost is the relevant concept.) The second is that the marginal cost of a good is proportional to the incremental amount of labor needed for its production. The theory of rent plays three roles in the classical theory. It is a theory of: (a) the sectoral allocation of labor between agriculture and manufacturing, (b) the distribution of income between land owners and laborers, and (c) the pricing of land.

In order to understand why Hicks and Samuelson eventually proposed the kind of foundations that they did propose for consumer theory, it is helpful to see clearly what Jevons was originally trying to accomplish. Jevons does not state very explicitly his goals in introducing a competing theory, or the considerations that lead him to choose a version of utility theory as the basis for that theory. These reasons can be reconstructed, though, from his prefatory remarks, from the selection of topics that he treats, and from the transparent and acknowledged analogy between utility theory and the theory of rent.

In the preface to the first edition of *The Theory of Political Economy* (1871), Jevons mentions the classical "wage-fund theory" as being logically circular, and part of Mill's theory of international trade as being both logically faulty and harmful as a basis for policy formulation (p. vi). The wage-fund theory lies outside the scope of his book, but he deals at some length with the question of welfare gains to international trade. In fact, it is the only specific policy issue that the book does examine. There is no indication, though, that his dissatisfaction with Mill's treatment of this issue had been the genesis of his rejection of the classical theory as a whole.

In fact, this passage in the preface can be read as a condemnation of the classical theory primarily for its logical incoherence rather than for the falsehood of its specific conclusions. On this reading, Jevons's aim is to construct a more cogent theory that encompasses as much as possible of the successful parts of the classical theory. The reading is consonant with further statements in the preface to the second edition, such as

> ...all branches and divisions of economic science must be pervaded by certain general principles. It is to the investigation of such principles—to the tracing out of the mechanics of self-interest and utility, that this essay has been devoted. The establishment of such a theory is a necessary preliminary to any definite drafting of the superstructure of the aggregate science. (*Ibid.* 1879, xvii-xviii)

Specifically, Jevons's skepticism is with regard to the labor theory of value. This is clear by elimination since he accepts the other component of the classical theory (i.e., the theory of rent) that is relevant to the part of economics that he treats. Four pieces of evidence can be found in the second edition of the book for this interpretation. Three of these can be dealt with quickly. First, Jevons discusses at length the ambiguity of the classical and popular use of the term "value," and he argues that the dimension of value ought to be zero rather than one (which would be its dimension if it were a quantity of labor, see [1957] 1965, 76-84). Second, he points out explicitly the shortcomings that labor would have as a general measure of value, even if it were true that relative prices could be explained by some notion of value (*ibid.*, 161-166). Third, as an application of his own theory, he accounts for the apparent success of its predecessor; that is, he explains why the labor theory of value often provides a good approximation to the relationship between relative price and labor inputs for a pair of commodities (*ibid.*, 189-193).

All of this evidence is consistent with the interpretation that the labor theory of value is only one among several major parts of the classical theory that Jevons is rejecting.[2] The fourth piece of evidence establishes that it is, in fact, principally this part of the classical theory that the new theory developed in Jevons's book is intended to replace. This evidence consists of the explanations of economic phenomena that Jevons provides via the theory. These explanations deal preponderantly with the levels and fluctuations of prices—the same phenomena that the labor theory of value is supposed to explain within the classical theory. Jevons provides a resolution of the diamond-water paradox (*ibid.*, 79-80), and he discusses why goods that are close substitutes in consumption have similar prices (*ibid.*, 134), why the prices of luxuries tend to be more stable than those of necessities (*ibid.*, 148) and why the price level of a staple commodity fluctuates more, relative to its mean, than does the supply of the commodity (*ibid.*, 152). One prominent piece of explanation that does not fit this pattern is the discussion of the relationship between the wage and an individual worker's labor supply, a discussion which serves another purpose that will be treated in a moment.

On this reading, Jevons's intent is to reformulate the classical theory in a fairly conservative way. He rejects the labor theory of value which is one of the most important components of that theory, but he does not by any means reject the theory as a whole. In fact, in the preface to the first edition, he proclaims this conservative approach by writing that "[h]ad Mr. Mill contented himself with asserting the unquestionable truth of the Laws of Supply and Demand, I should have agreed with him" (*ibid.*, vi). That is, he is in agreement with the classical tradition regarding what are the problems that economic theory is supposed to address, and he accepts a large part of the classical theory itself. His primary task is to reconstruct one specific part of that theory, the theory of price determination, for which he regards the labor theory of value as being an incoherent foundation. An ideal way to provide such a conservative reconstruction would be to formulate a theory of price determination that would be logically parallel to the theory of rent, and this is in fact what Jevons does. This strategy is particularly visible in a passage where he considers a consumer who must decide at the margin whether to retain money or to use it to purchase a consumption good (*ibid.*, 113-114). The quantity that would now be called "marginal utility" of money is, as a simplifying approximation, considered to be constant. ("Marginal utility" is the currently used name of the concept that Jevons called the "final degree of utility.") The marginal utility of the consumption good is considered to be decreasing. Thus, if the product of labor at the extensive margin[3] in agriculture is replaced by the marginal utility of the consumption good, the marginal product of labor in manufacturing is replaced by the marginal utility of money, and the price ratio between the agricultural and manufactured goods is considered to be unity, then Ricardo's simplest version of the theory of rent is formally transformed exactly into the particular version of the consumer's decision problem that Jevons is considering.

Jevons treats this version of the theory only as a special case of his more general theory, even though he notes that the special case is all that is "needed to represent the conditions of a large part of our purchases" (*ibid.*, 113). Thus it might be thought that the analogy with rent is no more than a happy accident. There is, however, a good reason why the special case is not completely adequate for Jevons's purposes. This reason has to do with one of the great advantages of his theory: that by treating labor as the part of the worker's endowment of time that is traded on the market, it subsumes the theory of labor supply under the theory of the consumer. However, according to the version of the theory that is exactly analogous to the theory of rent, an increase in the wage would engender an increase in the amount of labor that a worker would choose to provide, just as according to the theory of rent, an increase in the price of food will engender an increase in the amount of land that is cultivated. This prediction clearly contradicts the history of increasing wages and decreasing hours of work throughout western Europe in the century preceding Jevons's work. In order to avoid it, he introduces the concept of the increasing marginal disutility of labor as a function of its duration, and he shows that this concept renders the theoretical slope of the labor-supply curve indeterminate (*ibid.*, 179-180). The general version of Jevons's theory of demand, in which the consumer decides between two commodities that may both have decreasing marginal utility, provides the flexibility that is required to be consistent with a negative slope. Thus, it seems a good reconstruction to suppose that Jevons would have first constructed the special case of the theory of exchange, using the theory of rent as a model, and would have generalized it subsequently.

In any event, Jevons assumes precisely the properties of utility that are necessary to carry through the formal analogy of the special case with the theory of rent. It is worth noticing that this set of assumptions is more parsimonious in at least one important respect than those of his contemporaries. In particular, Jevons explicitly avoids making any commitment to interpersonal comparability of preferences, and he also believes that he can avoid, or at least assume only a modest form of, cardinal intrapersonal comparison of preference intensities (*ibid.*, 12-14). Actually, it is not clear that this latter claim can be meaningful. What Jevons claims is that "the reader who carefully criticises the following theory will find that it seldom involves the comparison of quantities of feeling differing much in amount" (*ibid.*, 13). But, unless Jevons is able to measure both large and small differences in quantities of feeling, how can he assure the reader that only small differences are involved? Moreover, Jevons can make this statement about the measurement of "feeling" only because he distinguishes between utility and preference. Utility is a relation holding between a consumer (or a consumer's preferences) and commodities (*ibid.*, 43). Utility must be cardinally measurable in order for the concept of diminishing marginal utility to make sense. At least, though, the discussion of "feeling" shows that Jevons has considerably more awareness and caution regarding the conceptual difficulties involved in formulating a theory of preference than the early neoclassical theorists are usually credited with today. When Jevons makes assumptions that fail to withstand subsequent philosophical scrutiny, it is because he believes that they are essential assumptions rather than because he is careless or obtuse.

Jevons makes three fundamental assumptions about the utility of commodities for any particular consumer. First, he assumes that utility is cardinally measurable by a real-valued function (*ibid.*, 45-52). Second, he assumes that this function is concave for each commodity, that is, that there is diminishing marginal utility (*ibid.*, 52-57). As has just been mentioned, this second assumption would entail that utility is cardinally measurable, even if Jevons had not explicitly adopted that assumption. Third, he assumes that the utility function is additively separable in distinct goods, in consumption of the same goods at different times, and with respect to incompatible random events.[4] That is, he assumes discounted temporal utility and expected utility under risk (*ibid.*, 71-73).

The first two assumptions are unquestionably necessary to formulate an analogue of the theory of rent. They correspond to the assumptions that the values of output in agriculture and manufacturing are cardinally measurable, and that the product of agricultural labor is decreasing at the extensive margin. In contrast, it is not apparent that the third assumption of additive separability is required to construct this central part of the theory, or to relate it to any of the applications that are given an extended treatment. Rather, this assumption seems to be made because it enables many plausible incidental conclusions to be drawn from the theory. This supposition is consonant with Jevons's remark that "Having no means of ascertaining numerically the variation of utility, [Daniel] Bernoulli had to make assumptions of an arbitrary kind, and was then able to obtain reasonable answers to many important questions" (*ibid.*, 159-160).

Before proceeding to discuss the status of these three assumptions, let us summarize what has already been established regarding the role and the substance of Jevons's theory of the consumer. Three important facts have been established. First, Jevons envisions this theory as providing a foundation for the theory of aggregate economic phenomena that is the main subject of classical economics. Second, he emphasizes the need to have a theory that is explicit, that is comprehensive, and that has a rigorous deductive and mathematical structure. (The most important sense in which Jevons's theory is comprehensive is that it provides a unified account of how the prices of produced goods, land, labor, and scarce goods in fixed supply are determined.) Third he provides such a theory, taking as axioms the three assumptions that have just been enumerated.

In emphasizing the need for an explicit, comprehensive, deductive theory, Jevons has already enunciated the first of the theses that characterize the Hicks-Samuelson program. With respect to the issue that the second of those theses concerns, the basis for acceptance of fundamental assumptions of the theory, Jevons is far from having a well articulated view. He actually comes close to opining that it is impossible to have good grounds for accepting the fundamental assumptions of any economic theory, writing that

[t]he final agreement of our inferences with *a posteriori* observations ratifies our method. But unfortunately this verification is often the least satisfactory part of the process, because, as J. S. Mill has fully explained, the circumstances of a nation are infinitely complicated, and we seldom get two or more instances which are comparable. (*Ibid.*, 18)

Nevertheless he does make numerous attempts to justify his three assumptions, and in doing so he suggests the possibility of basing their acceptance on four distinct kinds of evidence.

First, he proposes that the verification of consequences of the theory regarding broad economic aggregates serves to confirm the assumptions of the theory. Specifically he suggests that the prosperity of England in a free-trade regime provides such confirmation "as far as, under complex circumstances, facts are capable of doing so" (*ibid.*, 19). Appeal to aggregate evidence fits well with Jevons's conception of the role of consumer theory being to constitute a theoretical foundation for aggregate economics. However, as the quotations just presented make clear, he is not sanguine about the prospects for strong confirmation by this means.

Second, he proposes that a marginal-utility function can in principle be estimated from data regarding the prices and consumption levels of particular goods, and that the assumptions can then be verified directly. This is closely analogous to the second thesis held by Hicks and Samuelson. Jevons proposes a mathematical method for reconstructing the consumer's marginal-utility function from demand observations (*ibid.*, 146-148). He takes inconsistent positions regarding whether data sufficient for such a task exists, expressing confidence at one point but stating shortly afterward that existing statistics would not be sufficiently complete or accurate (*ibid.*, 10-11, 21). It is important to note that he has aggregate data, rather than individual-level data, in mind. Thus he implicitly assumes that aggregate data can be regarded as having been generated by a fictitious "representative consumer." (Subsequently it has been shown both theoretically and empirically that this assumption is not warranted in general. This issue will be discussed later in the paper.) Jevons expresses doubt that individual-level data would be useful even if it were available, pointing out that the theory cannot be hoped to be a very accurate description of the behavior of any individual, but that individual consumers' deviations from its predictions can reasonably be expected to cancel out on average (*ibid.*, 89-90).

Jevons's proposed method for reconstructing a marginal-utility function would presume that utility is cardinal. This is a matter of definition since marginal quantities are defined in terms of the magnitudes of differences of values. Thus, at best the consumption data just described could only serve to confirm the assumptions of concavity and additive separability of the utility function, but not its status as a cardinal representation. Jevons proposes a third kind of evidence, experimental data from physiology and psychology, to confirm the assumption of cardinality (*ibid.*, 35, 203-209).

Finally, Jevons suggests the usefulness of a fourth type of evidence, which in the "Introduction" he calls "intuition" (*ibid.*, 18). At that point he proclaims:

> [t]hat every person will choose the greater apparent good; that human wants are more or less quickly satiated; that prolonged labor becomes more and more painful, are a few of the simple inductions on which we can proceed to reason intuitively with great confidence. From these axioms we can deduce the laws of supply and demand, the laws of that difficult conception, value, and all the intricate results of commerce, so far as the data are available. (*Ibid.*)

However, he immediately qualifies this statement by cautioning that "[t]he final agreement of our inferences with a posteriori observations ratifies our method" (*ibid.*). Because the doubt regarding this evidential claim was the main motivation for Samuelson's later work, we will digress here to put the claim in its historical perspective.

When Jevons attempts to justify various claims on the basis of "intuition," he apparently is trying to use a method generally known as "introspection." Introspection is an attempt by a person to reflect on his or her mental processes and then to generalize these findings to all humans. In the mid-nineteenth century, Hermann von Helmholtz had made a serious attempt to develop introspection as a method of controlled scientific experimentation (Hearnshaw 1987, 132-133). However, by the end of the century there was a mounting skepticism towards this introspective method (Hearnshaw 1987, 290). Thus, if Jevons had really been trying to appeal to introspective evidence, that attempt would have been suspect. However, this appeal seems rather to have been Jevons's way of formulating a methodological principle rather than of recommending the use of a type of evidence.

Looking again at Jevons's statement concerning "intuition," we can see that, although he attempts to appeal to introspection as a method for obtaining evidence, in fact the examples he gives are not arrived at in this manner. His first two propositions concerning the "satiation of wants" and the choosing of the "greater apparent good" actually seem to be true by definition or stipulation, if they are true at all (Is hunger "more or less quickly satiated" during a famine?). Jevons's third proposition concerning the marginal disutility of labor also cannot be described as being obtained through introspection. Generalizations arrived at through introspection, by definition, should have few if any exceptions. By his own admission, however, Jevons's third proposition tends not to hold for self-employed professionals, a group which, in Jevons's time as well as our own, comprised a substantial portion of the working population ([1957] 1965, 181-182). Considering the size and importance of this "excepted" group, we must conclude that Jevons was not, in fact, using introspection. Although his proposition seems to possess some rough-and-ready explanatory force within a loose web of commonsense concepts, it is not the kind of statement one would want to say holds for everyone, knowing that so many people would not fall within its scope. Jevons's use of "intuitive" evidence seems instead to reflect the use of some sort of "folk psychology" in his reasoning, subject to the constraint that the results of that reasoning should ultimately be verified.

III. PARETO'S RESTATEMENT OF THE THEORY

In the three decades following the publication of the second edition of Jevons's *The Theory of Political Economy*, conceptual criticism of the foundations of consumer demand theory led to revisions of the formal theory and to the formulation of new theoretical research problems. The clear recognition by Irving Fisher (1892) that only the ordinal properties of the consumer's preferences are needed for the characterization of the consumer's optimizing behavior in the market, and Fisher's consequent insistence that the data generated by that behavior cannot provide a basis for determining the cardinal properties of the utility function, is the instance of such criticism that is most relevant to Jevons's particular formulation of the theory. We have seen that this insistence is not directly contradictory to Jevons's views, because Jevons does not explicitly claim that consumer-choice data alone are sufficient to confirm the theory and because he explicitly recognizes the usefulness of psychological and physiological experiments to generate further data that would be relevant to cardinality. Nevertheless, Fisher's criticism is important for two reasons. First, it establishes the dispensability of cardinality for consumer theory, if the role of the theory is simply to serve as a theoretical foundation for the explanation and prediction of aggregate economic phenomena. Second, it makes clear the insufficient clarity of Jevons's discussion of the dispensability of preference cardinality for his theory. According to Fisher, consumer-choice data can provide absolutely no information about the cardinality of preference intensities. If this argument is sound, then Jevons's assurance that the empirical interpretation of his theory does not require the accurate cardinal measurement of intense preference (as opposed to mild preferences) is no assurance at all.

In the first edition of the *Manual of Political Economy* (1906), Vilfredo Pareto addresses this second issue.[5] In the case of a consumer who chooses between two goods, Pareto presents a mathematically rigorous construction of indifference curves from a complete specification (at every possible consumption bundle) of indifference to infinitesimal changes of consumption, and he explains the equivalence of the family of these curves to the concept of an ordinal utility function ("index of ophelimity" in his terminology). In a review of this book later in 1906, Vito Volterra points out the invalidity of Pareto's construction in the general case where there are more than two goods. A second edition of the *Manual*, published in 1909, contains a mathematical appendix where the general case of this "integrability problem" is carefully treated. This second edition provides a revealing portrait of consumer theory as it had been developed prior to the work of Hicks and Allen. Of particular interest are the treatment of cardinality and the specification of criteria for acceptance of theoretical assumptions.

Pareto's views of these two issues are connected by his general views about the character and epistemic status of scientific theories, a topic to which he devotes the introductory chapter of the *Manual*. Pareto holds a correspondence theory of scientific truth, and he also envisions a theory as a statement of lawlike uniformities from which predictions can be deduced. He recognizes the tension between these two positions, and resolves it by proposing a methodology of scientific research programs. He writes that

[s]ince we do not know any concrete phenomenon completely, our theories about these phenomena are only approximations. ...Hence we should never judge the value of a theory by investigating whether it deviates in some way from reality because no theory withstands or will ever withstand that test....

It is necessary to substitute quantitative study for qualitative study and to investigate the extent to which the theory departs from reality. Of two theories we will choose the one which departs from it the least. We will never forget that a theory should only be accepted provisionally. ...Science is in perpetual development. (1906, part 1, 11)

He also takes a theoretical pluralist position, writing that

[A]bstraction is the result of subjective necessities... .

[T]herefore it is arbitrary, at least within certain limits, because the purpose which the abstraction has to serve must be taken into account. Consequently a certain abstraction...does not exclude another abstraction... Both can be used depending on the purpose one has in mind. (*Ibid.*, 21)

Pareto's discussions of specific assumptions of economic theory take these general principles as a background. For example, he freely admits that people do not know their preferences fully and that people's preferences are mutable, but argues that it would not be fruitful at the time he writes for economic theory to take these facts into account (*ibid.*, part 4, 26-28). This seems to be his view also regarding cardinality of utility. He states unequivocally that there really do exist intrapersonal variations in the intensity of preferences, but he endorses Fisher's position that the data of market behavior do not generally furnish a basis for measuring these variations (*ibid.*, part 3, 32-36). We will examine carefully below his belief that the case of additively separable utility constitutes an exception to this generalization. He notes that the convexity of the consumer's indifference curves to the axes of Euclidean space (where each distinct good defines an axis, and consumption possibilities are plotted with respect to this coordinate system) is a consequence of diminishing marginal utility, and that this ordinal assumption is sufficient to make the arguments for which Jevons had invoked diminishing marginal utility.

Pareto formulates a version of consumer theory without assuming cardinal utility, but he does not fully justify his choice of formulation. That is, he explains the feasibility of dispensing with cardinality, but he does not explain the desirability of doing so. We must impute an argument to him, then, on the basis of our best available understanding of what his considerations might have been. As a starting point, we would suggest that Pareto is implicitly in agreement with Jevons's views on the grounds for accepting an assumption, although it is evident from the passages quoted above that he has a somewhat different idea from Jevons of what degree of commitment such acceptance implies. The evidence for this interpretation is admittedly indirect. Pareto expresses respect for Jevons's work (*ibid.*, 30). He agrees with Jevons regarding the general principle that theories "must be in accord with the facts" (*ibid.*, part 1, 11), but he avoids entirely the issue of what specific kinds of statement count as being factual. He refers throughout the manual to instances of all four of types of proposition that Jevons regards as providing evidence for assumptions (e.g., *ibid.*, part 9, 23-26; 88; part 4, 33; 12 resp.). On at least one point, the need to test the assumptions of consumer theory against "collective and average" phenomena rather than against completely disaggregated data, he echoes Jevons's discussion of a specific issue regarding the evidential support for consumer theory (*ibid.*, part 4, 35). In summary, although he writes at length about methodological issues, and although he has ample opportunity to take issue with Jevons's views regarding evidence, he never does so. Although he never explicitly endorses those views, they would serve well to account for his own use of evidence.

In connection with the role of additive separability in Jevons's theory, we have seen that there are two kinds of consideration that come into play when an assumption is entertained. First, can the assumption contribute enough to the explanatory success of the theory to warrant bearing the increased evidential burden that it will impose on the theory? Second, can the assumption relate the theory to new evidence that will support the theory as a whole? Pareto's argument for the dispensability of cardinality constitutes, in effect, a negative answer to the first question. Pareto also supplies a negative answer to the second question (*ibid.*, 33). Namely, he has argued earlier (*ibid.*, 3) that the utility function must have as arguments the amounts of goods possessed rather than the amounts actually consumed, and he points out that the psychological and physiological experiments to support the cardinality assumption refer to quantities of actual consumption. He concludes that this evidence can provide only weak support, at best, for the hypothesis of cardinal utility.

Recall that, at the end of the preceding section, we have argued that reference to "intuitive evidence" is actually a disguised appeal to a methodological principle, rather than being truly an evidential claim. Specifically, this covert principle is that propositions drawn on "folk-psychological" utility theory should be granted temporary status as scientific hypotheses while evidence is sought to support them. Thus, we have argued that Jevons has only displayed three categories of possible evidence for consumer theory rather than four as he has supposed. Now the argument of Pareto that we have just discussed impeaches another of Jevons's categories of evidence.

If both Pareto's argument and our argument are accepted, then only Jevons's first two categories of evidence can be recognized. Recall that the first of these categories is the behavior of broad economic aggregates, and that the second is data regarding the prices and consumption levels of particular goods. We have already noted Jevons's assessment that evidence of the first category can be related only indirectly to consumer theory, and that therefore such evidence is not likely to be decisive. In view of this assessment, it would be reasonable to develop consumer theory in such a way that inference from aggregate evidence would play only an ancillary role, and that the primary burden of supporting the theory would be borne by the second category of relatively disaggregated evidence. This program is essentially the one that Slutsky, Hicks, and Samuelson adopted.

Before we proceed to consider how Hicks and Samuelson carry out this program, one complication has to be introduced. This complication concerns the scope of the class of "hard-core" theoretical assumptions that are exempt from confrontation with data. While Pareto does not refer explicitly to such assumptions in his discussion of scientific research programs, we contend that his consumer theory incorporates such a class, and that the class is defined in a precise way.

Note first that "hard-core" assumptions fit easily into Pareto's account of science. Strictly speaking, he takes whole theories rather than individual propositions to be the objects of acceptance or rejection. Thus there is no requirement that every assumption of a theory must be testable. Furthermore, Pareto views theoretical assumptions generally as describing "extreme cases" (*ibid.*, 58) and as admitting of exceptions. As a theoretical pluralist, he is in a position to recommend the formulation of alternative theories that dispose of a problematic assumption, while not necessarily recommending the excision of the assumption from the theory of which it originally forms a part.

Pareto's treatment of the integrability problem for utility makes it clear that he regards integrability (that is, the existence of a well-defined ordinal utility function that completely specifies the consumer's ordinal preferences regarding infinitesimal changes of consumption) as a "hard-core" assumption. He likens the status of ordinal utility to that of force and energy in Newtonian physics (*ibid.*, appendix, 7). He describes the establishment of mathematically sufficient conditions for an ordinal utility function to exist as being "a digression, not at all necessary in order to establish the theory of economic equilibrium, and...even outside it" (*ibid.*). Regarding the problem of verifying such conditions, he writes that

> [t]he fairly great difficulty, the impossibility even, that may be found in carrying out these experiments in practice, is of little importance; their theoretical possibility alone is enough to prove, in the cases we have examined, the existence of the indices of ophelimity, and to reveal certain of their characteristics. (*Ibid.*, 42)

In view of the subsequent work of Debreu, we can say that Pareto is willing to assume, even without prospect of eventually receiving empirical support for, the consistency properties of completeness and transitivity (and an Archimedean property) for ordinal preferences.

In contrast, Pareto clearly regards the question of whether utility can be expressed as an additively separable function of the goods consumed (which question he takes to be equivalent to the question of whether cardinality of utility is an economically meaningful concept) as being an empirical question (*ibid.*, 9-11). Pareto does not discuss whether the convexity of indifference curves to the axes is a "hard-core" assumption or one for which empirical support is to be demanded. We will see that the theoretical status of convexity is a question that is to cause serious trouble, and the answer to which the revealed-preference theory is ultimately to clarify.

IV. HICKS' S PROBLEM OF EVIDENCE

Unlike some of his predecessors, John Hicks wishes very much to rid his utility concept of casual evidence based on introspection. Especially crucial for Hicks is the removal of quantitative or measurable "utilities" from economic thought. At the beginning of *Value and Capital* (1946), he states: "We have now to undertake a purge, rejecting all concepts which are tainted by quantitative utility, and replacing them, so far as they need to be replaced, by concepts which have no such implication" (p. 19). In beginning this purge, Hicks accepts the notion discovered by Pareto, that the one sufficient component for determining an individual's purchases at given prices is a "given scale of preference" (*ibid.*, 18). We need only know that an individual prefers one set of commodities over another; we do not need to determine by what *degree* this preference occurs. This scale determines, in turn, the consumer's indifference map. Constructing utility theory on this basis, according to Hicks, allows us to use Occam's razor to cut out unnecessary assumptions from our economics.

Because they rely directly upon a quantitative notion of utility, the first two casualties of Hicks's purge must be the concept of marginal utility and the principle of diminishing marginal utility. These are replaced in the new theory by the concepts of "marginal rate of substitution" and "diminishing marginal rate of substitution," ideas which, according to Hicks, rely only upon ordinal preferences. He defines the marginal rate of substitution of x for y as "the quantity of y which would just compensate the consumer for the loss of a marginal unit of x" (*ibid.*, 20; variables changed to our notation). Using this definition, he gives the condition for an individual's equilibrium with respect to a set of market prices. This equilibrium occurs when the ratio of the marginal rates of substitution of two goods equals the ratio of their prices. The condition for equilibrium closely resembles Marshall's; the major difference between the two principles lies in Hicks's use of "marginal rates of substitution" instead of Marshall's "marginal utilities."

Like the notion of marginal rate of substitution, the principle of diminishing marginal rate of substitution is used in Hicks's theory in much the same way that Marshall uses diminishing marginal utility. That is, the principles are required for both authors if there is to exist a stable point of equilibrium between prices and income for the individual. The principles themselves, however, exhibit major differences. In particular, DMU allows for the case where, given two goods, increasing the amount of good x would reduce the marginal utility of y by more than that of x. If we look at the indifference curves for this case, we find that moving along the curve to the right actually increases the curve's slope. This case is not allowed for DMRS, for by this principle the indifference curves must be convex to the axes. In economic terms, DMRS states that "the more x is substituted for y the less will be the marginal rate of substitution of x for y" (*ibid.*, 21). DMRS must hold for an equilibrium point to be stable, for if the MRS is increasing at a particular point, it would prove advantageous for the consumer to acquire still more goods (and therefore consume on a higher indifference curve). The point would not, then, be one of equilibrium.

Hicks states that because we know from experience the existence of possible points of equilibrium on the indifference maps of nearly all individuals, we know that DMRS must sometimes be true. "However, for us to make progress in economics, it is not enough for us that the principle should be true sometimes; we require a more general validity than that" (*ibid.*, 22). In other words, Hicks wants to demonstrate the *law-like* character of DMRS; if it does not hold absolutely true, it should at least hold for the large majority of cases encountered.

In order to make DMRS more "generally valid," Hicks reflects upon the reasons for requiring such a principle. The rule should, according to Hicks, be "deduced from laws of market conduct" (*ibid.*, 23) which demonstrate alterations in consumer's behavior as conditions in the market change. According to these laws, these changes will result in the consumer's movement from one point of equilibrium to another. Clearly DMRS must hold at each of these points in order for the individuals to be in equilibrium at all. This does not, however, make DMRS a law. In order for this transition to occur, we must assume that DMRS holds at *all points between* those of equilibrium. Hicks makes this assumption by saying that there must be no "kinks in the curves" (*ibid.*) between these points. "If there are kinks in the curves, curious consequences follow, such that there will be some systems of prices at which the consumer will be unable to choose between two different ways of spending his income" (*ibid.*). In mathematical terms, the indifference curves of the individual must display convexity to the axes in order to be tangent to the budget line at only one point. We assume DMRS, then, in order to rule out such indecisive cases, and to provide us with the simplest possible theory.

But why does "simplicity" make DMRS law-like? Hicks elaborates further on his justification in two ways. First, he states that certain a priori propositions are arrived at in economic theory by assuming that, for a particular system of prices, there exists enough "regularity in the system of wants" (*ibid.*) for any set of quantities around those we are concerned with to possibly be a position of equilibrium. Second, he states that this assumption of regularity has in some way been *tested*—and that "its accordance with experience seems definitely good" (*ibid.*, 24).

By using this "regularity" assumption as the simplest basis on which to adopt DMRS, Hicks has produced several difficulties concerning the method by which this assumption may be adopted. In describing these difficulties, we will first of all show that they do *not* arise from Hicks's proposing a wholly new notion of theoretical simplicity as the crucial criteria for the adoption of economic premises. We argue, in fact, that the concept of "simplicity" acts merely as a smoke screen for Hicks, for there exists nothing in his work which would even remotely suggest the use of "simplicity" as a novel methodological principle. Second, we will show how Hicks appeals to both of our specified criteria for the adoption of assumptions concerning consumers' preferences, and how he violates each of these criteria. Finally, we demonstrate why Hicks encounters still greater difficulties because he wishes to use both of these criteria in justifying his assumption.

We have two reasons for believing that Hicks never seriously considered either defining or redefining simplicity criteria for the adoption of economic premises. First of all, he never makes statements implying that either Jevons or Pareto, his immediate predecessors, were misguided in their construction of criteria for accepting hypotheses. It would prove surprising if Hicks made this departure without making his opinions known to the public. Second, both Jevons and Pareto devote substantial space to methodological considerations. If Hicks had intended to make any new points with respect to this topic—whether in agreement with these men or not—one would think that he, too, would devote some space to this type of discussion. Yet he does not make this type of commentary; he merely states simplicity as a criteria for the acceptance of an assumption. We must therefore conclude that Hicks has no new ideas concerning simplicity as acceptance criteria, and that he follows his predecessors in using the two acceptance criteria (i.e., that consistency assumptions need not be adopted in the light of evidence, but that assumptions placing substantive restrictions on patterns of consumers' choices do require empirical support) we have previously identified.

Hicks does not speak of the criteria in these words but proposes discussion in terms of the "regularity" of consumer behavior. We want to argue that at various points he indicates both that "regularity" is a matter of "consistency" and of "substantive restrictions," and that he cannot decide under which category "regularity" should fall.

Hicks does not give an explicit definition of "regularity," but two conditions are clearly recognized as necessary conditions of regularity. First, a consumer's behavior is "irregular" if the consumer is not able to identify the unique choice that is optimal. Second, a consumer's behavior is "irregular" if her or his demand is not appropriately continuous as a function of prices. We are going to consider these two specific conditions of regularity with respect to each of the acceptance criteria that Jevons proposes. We argue that this sort of regularity cannot plausibly be subsumed under "mere consistency" and that Hicks lacks the evidence that the acceptance criteria for "substantive restrictions" requires.

In order to show that neither necessary condition of regularity can be coextensive or synonymous with consistency, let us consider two examples with two different types of heating oil. In both of these examples, the consumer clearly satisfies the consistency axioms, but does not satisfy the necessary conditions of regularity. We attach nothing more than cost minimization to the idea of consistency in these examples. First, let us consider the case where a person must choose between two different types of heating oil: one, the "low grade" brand; the other, the "high grade" brand. We assume that while the low grade brand costs half as much as does the high grade brand, the high grade brand burns twice as efficiently as does the low grade. Consumers thus cannot satisfy Hicks's first condition; they have no way of choosing between ways of spending their income. There is nothing inconsistent about this situation, but it is irregular according to the first way Hicks fleshes out the concept.

In our second example we demonstrate how the consumer might satisfy the consistency axioms yet not satisfy the second necessary condition of "regularity." First, let us imagine that on our consumer's furnace there is now placed a switch, one which controls which fuel may be used (only one fuel may be used at a time). Now, this switch may only be turned from one position to another by qualified personnel, who currently charge $40 per visit. Let us assume that, in order to heat her or his home for the winter, the consumer requires either one hundred gallons of low grade fuel or fifty gallons of high grade fuel (recall that the high grade fuel is twice as efficient as the low grade). Currently, low grade fuel costs $1 per gallon, while high grade costs $2.01 per gallon. Because it is cheaper to heat the home with low grade fuel, the consumer (whose furnace switch is already set on "low grade") will initially choose to use low grade. Now let us say that, before she or he actually goes out and purchases the fuel, the price of high grade fuel drops to $61 for fifty gallons. Although the price for high grade has fallen dramatically, changing to high grade would cost both $61 and an additional $40 to change the setting on the furnace. The consumer would therefore behave consistently in saving $1 and staying with the low grade.

If, however, during the same period in which the consumer is making this decision, the price of high grade drops just a bit more, say, to $59 a gallon, it would cost only $99 ($59 for the fuel and $40 to change the setting) to switch to high grade. The consumer would now behave consistently in purchasing the high grade. In fact, she or he would behave consistently in purchasing the high grade at any price below $60. Thus, by changing the price of the high grade fuel by an infinitesimal amount, we have changed its demand from zero to some large positive amount. While behaving consistently, then, the consumer again behaves "irregularly" because her or his demand is globally discontinuous as a function of prices.

Granting the correctness of these two examples, we have demonstrated the falsity of Hicks's apparent view that consistency in some way implies regularity.

If Hicks's regularity assumption is not an example of our consistency criteria, then we must look on it as a restriction of consumers' choice by our "substantive restrictions" criteria. Hicks therefore needs to show that his assumption can be justified using evidence from consumer demand. We can, however, find no empirical evidence at all to support Hicks's assumption at the time he was writing (1946). We would strongly argue, in fact, that the computational power required to conduct a test of this type of assumption has not been available until the last few decades, and, even when available, tests aimed at confirming assumptions similar to Hicks's do not offer nearly the kind of support Hicks claims to have had for his assumption several decades earlier (see Deaton and Muellbauer 1980, 67-73; they show how little evidence exists for the existence of, e.g., homogeneous demand functions). Even when considered to be a restriction on how consumers make their choices, it is far from clear how Hicks's assumption may be justified.

Given the absurdity of Hicks's claims to empirical evidence, we must ask why he names this claim in the first place. We would like to offer two alternative scenarios which may help us answer this question. On the one hand, we might say that Hicks's knowing that the truth of the regularity assumption would establish DMRS (convexity) as a law and would provide him with simple, plausible demand functions, simply alleged the existence of data he did not have, and, in effect, made statements for which he had no empirical support. On the other hand, although he must have kept in mind the justification of DMRS, he may have looked to alternative sources for evidence. In particular, he may have looked to his own preferences, believed that they conformed to the regularity assumption, and applied his conclusions to all consumers. In other words, the "evidence" Hicks discusses may have been introspective in character. Considering the period in which Hicks writes *Value and Capital*, however, we know that at this time introspective claims had little, if any, evidential status at all. Why Hicks would choose to use a method for obtaining evidence which clearly had no friends in the science of his time is, to say the least, unclear.

What seems to us most damaging to the regularity assumption, more so than either failed attempt at justification, is the fact that Hicks simply cannot decide *how* he wishes to make this justification. He cannot have it both ways. He must either accept the condition of regularity as an axiom, unsupported (and without need of support) by evidence, or he must take it as restricting the kinds of choices the consumer can make, and support it empirically. He attempts to use it in both ways, but, as we have demonstrated, he cannot use it in either way.

V. SAMUELSON: REVEALED PREFERENCE AND UTILITY

Although we attribute the research program we are currently studying to Hicks and Samuelson, it is not at all clear that in 1938 one would have wanted to make this claim. This is due to the fact that in an article of that year, "A Note on the Pure Theory of Consumer's Behavior" (1938), Samuelson comes out dead against using what he calls "the last vestiges of the utility analysis" (p. 61) in economics assumptions. Among these remnants he identifies the notion of DMRS as characterized by Hicks and Allen. "Why should one believe in [DMRS] except insofar as it leads to the type of demand functions in the market which seem plausible" (*ibid.*, 61). He claims, as we do, that Hicks's formulation of DMRS contains introspective elements, and that without these elements, the concept seems unclear in meaning and usage.

In order to rid economics of such unsightliness, Samuelson proposes that we start anew, using a "different set of postulates" (*ibid.*, 62) on which to describe consumer behavior. He assumes three postulates. As his first postulate, he assumes the existence of n commodities, each a fraction of the (given) prices of all commodities and the consumer's income. This set of equation is subject to the budget constraint, giving us $n + 1$ equations in n unknowns. Postulate 2 states that our functions are homogeneous of degree zero, allowing us to set the first price as numeraire. This postulate also makes the assumption that for two batches of commodities X and X' at price-income situation P,

$$[X^b P^a] \leq [X^a P^a]$$

implies

$$X^b < X^a$$

where $[X^b P^a] = \Sigma_{i=1}^n X_i^b P_i^a$, $[X^a P^a] = \Sigma_{i=1}^n X_i^a P_i^a$, and $X^b < X^a$

implies that batch one was selected over batch two. Analogously,

$[X^a P^b] \leq [X^b P^b]$ implies $X^a < X^b$ and, logically, $X^a \nless X^b$ implies $[X^a P^b] > [X^b P^b]$.

The third, and most important postulate has become known as the weak axiom of revealed preference or WARP. It assumes the consistency of the individual's choice behavior by stating that

$$X^b < X^a$$

implies

$$X^a \nless X^b$$

"In words this means that if an individual selects batch [a] over batch [b] he does not at the same time select [b] over [a]" (*ibid.*, 65). This statement, however, is not intended to have a purely tautological meaning; later uses of the postulate indicate that it should be defined over two periods of time. That is, if the consumer selects batch [a] over batch [b] when batch [b] does not cost more than [a], then whenever batch [b] is selected the consumer cannot afford batch [a].

Using these three postulates, taking the demand function as given and assuming that consumers spend all of their income, Samuelson is able to derive the semidefiniteness of the substitution matrix, indicating his ability to arrive at the basic results of ordinal utility theory without bringing in questionable concepts such as DMRS.

We should make two notes at this point. First of all, Samuelson (1938) expresses almost no interest in the problem of integrability, except to state that "I should strongly deny...that for a rational and consistent individual integrability is implied, except possibly as a matter of circular definition" (p. 68). We will soon see that this problem takes on a far greater importance than he would have imagined. Second, Samuelson certainly does not wish to deny the *results* of ordinal utility theory; he merely wishes to offer clearer and less objectionable foundations for deriving these results.

Samuelson's 1948 paper "Consumption Theory in Terms of Revealed Preference" provides another very important result: the construction of an individual's indifference map using the theory of revealed preference. Assuming that for two goods x and y, only one price ratio may be given to any one combination of their quantities, Samuelson writes the equation $-P_x / P_y = -f(x,y)$; that is, he gives the price ratio of two goods as a function of the quantities of each of the goods. The function f is assumed to be observable, continuous, and to have continuous partial derivatives. Now, if we draw a two-dimensional graph identifying the axes with the amounts of quantities x and y, any point will have associated with it a budget line on which it is revealed to be preferred to any other pair of quantities on or below that line. The slope of this line is given by the price ratio $-P_x / P_y$. If we then identify a slope dy/dx with this price ratio, we have, from the previous equation $dy/dx = -f(x,y)$.

Samuelson recognizes the fact that this simple differential equation has as a solution a family of curves in the (x,y) plane. He therefore uses two processes, that of Cauchy and Lipschitz and one of his own invention, in order to approximate this solution. He also shows that when approximating a solution point A from below, all points except those on the frontier of A are revealed to be inferior to A. A similar approximation from above shows all points except those on the frontier are revealed to be superior to A. Those points "lying literally on a (concave) frontier locus" (*ibid.*, 251) can be revealed to be neither inferior nor superior to A, and may therefore be spoken of as being indifferent to A. Using all points on the (x,y) axis Samuelson is able to construct a complete indifference surface for this plane.

While making this construction, Samuelson also clarifies the role and importance of the convexity of the indifference curves to the axes. For, as Samuelson states, each point on the (x,y) plane has associated with it a budget line convex to the axes, and by the rules of revealed preference, we can rule out the case where points on two or more convex sets form a concave set. We make our construction, then, by combining these budget lines into sets which preserve their convexity, even in the limiting case of the indifference curve.

At the end of this paper (1948), Samuelson in the text states that "the whole theory of consumer's behavior can thus be based upon operationally meaningful foundations in terms of revealed preference" (p. 251). The footnote attached to this statement, however, is not so bold:

> The above remarks apply without qualification to two dimensional problems where the problem of 'integrability' cannot appear. In the multidimensional case there still remain some problems, awaiting a solution for more than a decade now. (*Ibid.*, n. 2)

It is apparent, then, that the problem of integrability has taken on a far greater importance for Samuelson, for without a solution to this problem the revealed preference approach to constructing indifference maps cannot be made general.

He does not, however, have to wait very long for a solution. In the May 1950 issue of *Economica*, H. S. Houthakker offers both a critique and a modification of WARP (Samuelson 1950 endorses Houthakker's approach). The major problem with WARP, according to Houthakker, is that while we can know, for example, that $P^aX^b < P^aX^a$ and that $P^bX^c < P^bX^b$, we have no guarantee that, for a system of prices P^c, that $P^cX^a < P^cX^c$. "This means that if X^b is revealed to be inferior to X^a and X^c is revealed to be inferior to X^b, then the hypothesis does not rule out the possibility that X^a is revealed to be inferior to X^c" (Houthakker 1950, 161). Obviously, then, Houthakker needs to modify WARP to include some sort of transitivity rule.

This modification results in SARP, or the strong axiom of revealed preference:

> If $X^0, X^1, X^2,...X^T$ is a sequence of batches of goods such that each batch is bought at prices $P^0, P^1, P^2,...P^T$ respectively, and if at least two of these batches are different, and if the cost P^tX^t of each batch X^t at prices P^{t-1} is not greater than the cost $P^{t-1}X^{t-1}$ of the preceding batch in the sequence X^{t-1} at the same prices, then the cost P^TX^T of the last batch X^T at prices P^T is less than the cost P^TX^0 of the first batch X^0 at the same prices.

Using this rule, Houthakker is able to derive WARP (just set $T = 1$), and, using a similar approximation process to Samuelson's, is able to construct indifference surfaces without facing the integrability problem at all. In fact, he shows that this problem arises only when assumptions are incomplete; although the rules of asymmetry and semitransitivity are used for the same purpose, only the latter gives a formal equivalence of ordinal utility theory to revealed preference theory for the general case.

VI. THE DIFFICULTY OF APPLYING THE THEORY

Today it is fifty years since Hicks and Samuelson introduced their ideas, and forty years since the the formal equivalence between their approaches began to be well understood. Ordinal utility maximization continues to be, by far, the most widely accepted foundation for an economic theory of consumers' choices and behavior. It has maintained this position although it has been challenged by at least one serious alternative candidate for this status; Herbert Simon's behavioral theory. It would seem that Hicks and Samuelson and their followers have carried the day.

It is certainly correct that the theory of ordinal utility maximization has carried the day, and that the formalization of the theory attained its mature form in their work, along perhaps with that of Debreu. This does not mean that development of the theory has ceased. For example, it has subsequently been shown by Hugo Sonnenschein (1971) and subsequent researchers that even the ordinal-consistency assumptions on preferences can be relaxed without completely depriving the theory of economic content. However, this and other results about consumers' optimization under very weak assumptions have been understood to be demonstrations of the robustness of the standard version of the theory, rather than to be in competition with it.

It is also true that researchers in this program have largely achieved what they had set out to accomplish. Their work was directed single-mindedly towards the formulation of a theory that could be tested with the use solely of price and demand-quantity data for particular commodities. By the mid 1960s, they had achieved two objectives that economists since Jevons had been attempting. First, they had formulated an algorithm that, from such data, would recover the economically relevant information about the preferences that had generated it. (Such a revealed-preference algorithm was formulated explicitly by S. Afriat. Second, they had shown that the assumption of convexity of indifference curves implies no additional restrictions on demand data once the ordinal-consistency assumptions have already been made. That is, any finite set of demand data can be explained as having been generated by the maximization of convex preferences if it can be explained as having been generated by the maximization of any preferences at all. Third, John von Neumann and Oskar Morgenstern (1947) had given an ordinal characterization of expected utility, and Gerard Debreu (1960) had characterized separable utility. More recently, Hal Varian (1983) has also succeeded in characterizing the restrictions that expected-utility maximization places on budget-constrained demand for risky financial assets.

However, the application of these methods to actual data has been much less extensive and influential that what some of these researchers would have expected, especially given the vast improvement of computing technology that has occurred during the past four decades. The treatment of additive separability is a case in point. The assumption of additive separability continues to have widespread and important use in theory, and it is frequently imposed when statistical models of consumer demand are estimated. Undoubtedly, additive separability is often rejected statistically as a description of the ordinal preference that generate specific data sets, so the ability to make this test does have some effect on how these data are analyzed. However, we are aware of only two instances in which concerted work has been done with the goal of resolving empirically a dispute in which separability is a serious theoretical issue (as opposed to being merely an ad hoc restriction imposed for computational or statistical convenience). The first instance is the work of several labor economists including Thomas Macurdy, Joseph Altonji, and John Ham concerning the issue of whether or not workers' utility is separable in leisure at distinct dates. The answer to this question is relevant to assessing how large a proportion of unemployment is involuntary. This investigation has been inconclusive so far, though (the work to which we refer is surveyed in Blundell 1987). The other instance is the current work of Larry Epstein and Stanley Zin (1987) and of Orazio Attanasio and Gugliermo Weber (1987), who are attempting to explain anomalies in the theories of the consumption-saving relationship and of financial-asset pricing by means of utility functions that violate the expected-utility hypothesis.

Thus the Hicks-Samuelson theory has been less important from a practical point of view than might have been expected. This scientific disappointment is partly accounted for by the discovery by Hicks (1956) that the aggregate demand by all consumers for a set of commodities may not satisfy the Weak Axiom of Revealed Preference, even if the demand function of each individual consumer satisfies the axiom. This discovery raises a serious question regarding the application of the theory of the individual consumer to aggregate data. This question is intensified by subsequent work of Hugo Sonnenschein, Erwin Diewert, and Rolf Mantel, showing that only the most trivial implications of utility maximization for individual demand (specifically that all income is spent and that doubling of all prices and incomes would have no effect on demand) can be asserted of the level and first derivatives of aggregate demand at a particular price and distribution of wealth (this result, as well as even more negative results about the restrictiveness of aggregate demand in a general-equilibrium setting, is discussed by Shafer and Sonnenschein 1982). These are results that were clearly unanticipated by Hicks and Samuelson, who had envisioned that the addition of individual's demands would preserve the restrictions that were satisfied by each. It has already been noted that the ability to aggregate demands has long been thought to be an important consideration in favor of the testability of consumer theory since individuals have been presumed typically to conform to the predictions too imperfectly to confirm it. Thus, these results have highly negative implications for the Hicks-Samuelson program. It goes without saying, though, that the results would never have been proved had the program not been pursued.

VI. THE ACHIEVEMENT OF CONFIRMATION
BY AGGREGATE PHENOMENA

A second reason for the relatively weak influence of the Hicks-Samuelson theory on economics as a whole has been the success in the 1950s and subsequently of consumer theory, incorporating separability assumptions, in accounting for phenomena at a broad aggregate level. As we pointed out early in this paper, these were the phenomena with which the nineteenth-century founders of neoclassical theory seemed to be ultimately concerned, but that they had presumed to be too complicated and idiosyncratic to be of much value for scientific confirmation. In view of these empirical successes, many economists today are much more willing than were their predecessors to treat the rational consumer as an idealization (analogous to the physical idealization of an object as a point mass) that need not be literally correct, and to evaluate the theory on the basis of its implications about aggregates. This viewpoint was enunciated by prominent economists such as Fritz Machlup (1955, 1960) and Milton Friedman (1953), sometimes in rather naive or extreme terms. These economists' formulations have been subject to a great deal of criticism as general philosophical positions, but it is more appropriate and informative to view them as constituting specific recommendations to students and colleagues regarding scientific practice (Hausman 1981a, 136-139, discusses succinctly some of the difficulties that beset Machlup's and Friedman's formulations, and in that book he articulates a more sophisticated version of their thesis). On this understanding, the viewpoint needs to be assessed in terms of the success of the scientific contributions to which it leads. Two of these successes have been particularly noteworthy. One is the theory of lifetime allocation of wealth, including the "permanent-income theory" of Friedman (1957) and the "life-cycle hypothesis" of Franco Modigliani and Richard Brumberg (1954). The other notable success is the "capital asset pricing model" developed by Harry Markowitz (1959), James Tobin (1958) and others, which has provided a fairly satisfactory account of the determinants of the relative prices of various financial assets to one another. We now give a brief description of each of these contributions, emphasizing how it can be viewed as providing an aggregate-level confirmation of consumer theory.

In the 1950s, the most widely accepted theory of the relationship between real income (i.e., income measured in a way that corrects for inflation) and consumption was the theory proposed by Keynes in *The General Theory*: that there is a stable relationship between income and consumption, that the absolute amount of consumption rises with income, but that the proportion of income consumed falls and saving rises as income increases. A number of empirical findings were inconsistent with that theory. For example, the proportion of income consumed had remained roughly constant throughout the period from 1900 to 1950, although real income had risen substantially during that period. Moreover, there was inconsistency between the measurement of this proportion as the slope of a time path of aggregate data *versus* other measurements based on a contemporaneous sample of households (these and other issues are discussed in chap. 4 of Friedman 1957). Friedman along with Modigliani and Brumberg applied the theory of the consumer in related ways to provide an alternative account of the income-consumption relationship, and they showed that their neoclassical accounts were consistent with the facts on which the Keynesian theory foundered. The essence of their argument is that consumption at different dates ought to be regarded as different goods, and that the access of a household to these goods is limited by a "lifetime budget constraint." That is, the household makes its consumption decisions on the basis of its best estimate of both its current and future earnings, rather than mechanically on the basis of its current earnings alone as Keynes had assumed. In particular, a household that has its income *temporarily* reduced will revise its estimate of its "lifetime budget constraint" downward, but this reduction will not be fully proportional to the temporary reduction because it is anticipated that the usual level of income will be restored. Consequently, if consumption is plotted as a function of current income alone and if households' changes in income are predominantly temporary ones, then this consumption function will have the Keynesian features. However, if a time series of data is used where long-term growth of income is much larger than the magnitude of temporary fluctuations, then a constant proportion of consumption to income will be seen. Thus, the income-consumption relationship derived from neoclassical theory succeeds in accounting for the several anomalies that faced the Keynesian theory. This success can be regarded as confirmation of the neoclassical theory.

The second major success of consumer theory has been the development of the capital asset pricing model, which provides a substantially improved account of the determinants of the relative prices of various financial assets to one another. In the 1950s it was well understood that the price of an asset would be positively related to its expected return and negatively related to its riskiness. This was called the "mean-variance theory." The expected return and riskiness of an asset were typically estimated from a historical time series of its price, interpreting the (logarithmic) time trend to measure the expected return and the standard deviation of the series from its trend to measure the riskiness. The relationship of expected return to price was well established, but there were known to be some assets that were priced high (controlling for expected return) although they had high measured riskiness.

In the 1950s and early 1960s, Markowitz, Tobin and others studied the portfolio-allocation problem of a consumer who maximizes expected utility. In this study, asset prices are supposed to vary across states of the world, and wealth in the different states of the world is interpreted as being different goods. Under some parametric assumptions about the utility function and the price distribution, expected utility of a portfolio can be expressed in terms of the composition of the portfolio and the first two moments of the joint probability distribution of asset prices. Among the second moments are the covariances of the price distributions of distinct assets, and an implication of the neoclassical theory is that market-clearing asset prices need to reflect these covariances. In particular, the covariances can account for the pricing of many assets that were anomalies for the mean-variance theory: the returns of these assets have high variance, but they have unusually low or even negative covariance with other assets that represent a substantial proportion of the wealth of the economy. Thus the owner of such an asset assumes one risk but hedges another, more significant, risk.

Like the theories of Friedman and of Modigliani and Brumberg, the test of this model relies upon aggregate data based primarily on the behavior of the financial markets. It requires no individual budget data, nor does it use the comparative-statics approach that economists from Pareto to Samuelson envisioned as the appropriate test of a theory of this type. Rather, the theory has implications regarding aggregate data that are as easily accessible as the financial pages of a daily newspaper. In fact, both of the applications of consumer theory discussed in this section seem to fulfill the hope of Jevons and other early neoclassical economists that the careful analysis of such routinely generated and accessible data would serve to confirm their theory.

VII. CONCLUSION

It may yet be a long time before a final verdict on the scientific usefulness of Hick's and Samuelson's theory can be given. The problem of applying the theory to aggregate data is severe, but is not necessarily intractable (Deaton and Muellbauer 1980, 214-379, contains a wealth of material concerning recent efforts to refine the theory and to test these refinements). Furthermore, although the empirical successes of the past four decades have been impressive, they are not conclusive. In particular, it has recently been suggested both that consumption data may be "too volatile" to be consistent with Friedman's theory, and that one of the principal theoretical constructs of the capital asset pricing model may lack any close empirical counterpart. (For a survey of a number of recent studies of consumption and savings behavior, see King 1985; probably the most influential criticism of the testability of the capital asset pricing model has been made by Roll 1977). Although it has seemed in the recent past that the testability of consumer theory need not hinge on revealed preference theory, it is not impossible that that assessment might be revised.

Furthermore, although the problem of relating a theory of individual behavior to aggregate data has been a severe one for the Hicks-Samuelson theory, that problem is neither unavoidable nor completely intractable. Since the 1960s, a number of large data sets concerning the consumption behavior of individual households have been compiled, and contemporary computing technology makes it feasible to analyze this data. There has also been some recent theoretical progress in using the theory to derive propositions about aggregate demand when certain assumptions about the distribution of wealth across households are satisfied (see Hildenbrand 1983). Thus, while the Hicks-Samuelson program has certainly receded in influence since the 1950s, it cannot yet be known whether that recession is temporary or permanent.

What can be said with certainty is that the research program of neoclassical consumer theory during the first half of this century, culminating in the work of Hicks and Samuelson, was a well conceived one. It was begun in order to relate the neoclassical theory of the consumer to the only body of data that seemed capable of providing a test. It achieved an impressive measure of success in characterizing the implications of the theory for that data, only to be frustrated at the end by the aggregation problem that had been unforeseen until an apparently workable theory was in existence. Almost contemporaneously, a kind of empirical work that had long been considered unpromising began suddenly to make enormous strides. These are events of the kind that make science an exciting and risky enterprise, and are not at all evidence that researchers were lacking in sophistication or good judgement. Indeed, regardless of the ultimate fate of the Hicks-Samuelson program, we should view this episode as being one of the landmark examples of scientific-theory construction in the history of economics.

Minneapolis, MN

Chicago, IL

Notes

*This paper was begun while Keith Moss held a fellowship in the Department of Economics at the University of Pittsburgh, sponsored by the Scaife Family Foundations. It is a pleasure to acknowledge the influence on the paper of discussions with Jean Hampton, Peter Machamer, J. E. McGuire, and Candace Vogler.

1. Samuelson had initially proposed revealed-preference theory as a more parsimonious alternative to ordinal utility theory. By 1950, though, he had endorsed the reasonableness of SARP, and the goal of research on revealed preference had shifted to the attainment of a deeper and more general understanding of the equivalence between the two formulations of the two theories. This new research focus is clearly evident in the papers published in Chipman and others (1971). Also at that time a third formalism, using the elementary theory of a binary preference relation, began to be studied intensively. The work of Debreu (1954) quickly established that this is equivalent to ordinal-utility theory, under very mild assumptions.

2. Clearly, if his theory is to be completely general, he must also reject the scarcity theory of relative prices that classical economists had introduced to account for the

pricing of goods not produced by labor (and to deal with the "diamond-water paradox"). This is an ad hoc "fix," though, rather than being a major part of the classical theory.

3. The term "extensive margin" refers to the employment of a unit of labor to cultivate a parcel of land that was previously undeveloped, as opposed to the employment of an additional unit of labor on land that is already under cultivation.

4. Additive separability of utility with respect to distinct goods is not asserted directly, but is implicit in Jevons's mathematical formalism ([1957] 1965, 99). It is possible that the use of this formalism reflects a desire to put the argument in the simplest possible terms rather than indicating any real commitment to its deductive implications. However, Jevons nowhere indicates that he is making an expository simplification, either. In the case of the constant marginal utility of money, where he is making such a simplification but does not want to be committed to its full theoretical implications, he does indicate what he is doing very clearly.

5. The *Manual* is best known for its exposition of social-welfare theory on the basis of unanimous preference, without the assumption of interpersonal comparability of utility.

References

Afriat, S. (1967), "The Construction of Utility Functions from Expenditure Data," *International Economic Review 8*: 67-77.

Attanasio, O. and Weber, G. (1989), "Intertemporal Substitution, Risk Aversion and the Euler Equation for Consumption: Evidence from Aggregate and Average Cohort Data London," *Economic Journal 99*: 59-73.

Blundell, R. (1987), "Econometric Approaches to the Specification of Life-cycle Labour Supply and Commodity Demand Behaviour," *Econometric Reviews 6*: 103-65.

Chipman, J. S. and others (eds.) (1971), *Preferences, Utility and Demand*. New York: Harcourt Brace Jovanovich.

Deaton, A. and Muellbauer, J. (1980), *Economics and Consumer Behavior*. Cambridge, England: Cambridge University Press.

Debreu, G. (1954), "Representation of a Preference Ordering by a Numerical Function," in R. Thrall, C. Coombs and R. Davis (eds.), *Decision Processes*. New York: John Wiley and Sons, pp. 159-65.

_____ . (1960), "Topological Methods in Cardinal Utility Theory," in K. Arrow, H. Karlin and H. Scarf (eds.), *Mathematical Methods in the Social Sciences*. Stanford: Stanford University Press, pp. 16-26.

Epstein, L. and Zin, S. (1989), "Substitution, Risk Aversion and the Temporal Behavior of Consumption and Asset Returns: A Theoretical Framework," *Econometrica 57*: 937-69.

Friedman, M. (1953), *Essays in Positive Economics*. Chicago: University of Chicago Press.

_____ . (1957), *A Theory of the Consumption Function*. Princeton: Princeton University Press.

Hausman, D. (1981a), *Capital, Profits, and Prices*. New York: Columbia University Press.

————. (1981b), "Are General Equilibrium Theories Explanatory?," in J. Pitt (ed.), *Philosophy in Economics*. Dordrecht: Reidel, pp. 17-32.

Hearnshaw, L. S. (1987), *Sources of Modern Psychology*. London and New York: Routledge & Kegan Paul.

Hicks, J. R. (1946), *Value and Capital*. Oxford: Oxford University Press.

————. (1956), *A Revision of Demand Theory*. Oxford: Oxford University Press.

Hicks, J. R. and Allen, R. D. G. (1934), "A Reconsideration of the Theory of Value," *Economica 14* (I): 52-76 and (II): 196-219.

Hildenbrand, W. (1983), "The Law of Demand," *Econometrica 51*: 997-1020.

Houthakker, H. S. (1950), "Revealed Preference and the Utility Function," *Economica N.S. 17*: 159-74.

Jevons, W. S. ([1957] 1965), *The Theory of Political Economy*. 5th ed. Reprint. New York: Augustus M. Kelley.

King, M. (1985), "The Economics of Saving: A survey of Recent Contributions," in K. J. Arrow and S. Honkopohja (eds.), *Frontiers of Economics*. Oxford: Basil Blackwell, pp. 227-94.

Machlup, F. (1955), "The Problem of Verification in Economics," *Southern Economics Journal 22*: 1-21.

————. (1960), "Operational Concepts and Mental Constructs in Model and Theory Formation," *Giornale degli Economisti 19*: 553-82.

Markowitz, H. (1959), *Portfolio Selection*. New York: John Wiley & Sons.

Modigliani, F. and Brumberg, R. (1954), "Utility Analysis and the Consumption Function: An Interpretation of Cross-section Data," in K. Kurihara (ed.), *Post-Keynesian Economics*. New Brunswick: Rutgers University Press, pp. 388-436.

Pareto, V. ([1909] 1971),*Manual of Political Economy*. Reprint. Translated by A. Schwier and edited by A. Schwier and A. Page. Originally published as *Manuel d'Economie Politique* (Paris: V. Giard et E. Briere) New York: Augustus M. Kelley.

Roll, R. (1977), "A Critique of Asset Pricing Theory's Tests: Part I: On Past and Potential Testability of the Theory," *Journal of Financial Economics 6*: 129-76.

Rosenberg, A. (1976), *Microeconomic Laws: A Philosophical Analysis*. Pittsburgh: University of Pittsburgh Press.

Samuelson, P. A. (1938), "A Note on the Pure Theory of Consumer's Behavior," *Economica 5*: pp. 61-73.

————. (1948), "Consumption Theory in Terms of Revealed Preference," *Economica 15*: 243-53.

————. (1950), "The Problem of Integrability in Utility Theory," *Economica 13*: 355-85.

Shafer, W. and Sonnenschein, H. (1982), "Market Demand and Excess-demand Functions," in K. Arrow and M. Intrilligator (eds.), *Handbook of Mathematical Economics*, vol. 2. Amsterdam: North Holland, pp. 671-93.

Slutsky, E. ([1915] 1952), "On the Theory of the Budget of the Consumer," in G. Stigler and K. Boulding (eds.), *Readings in Price Theory*. Reprint. Translated. Originally published as "Sulla Theoria del Bilancio del Consumatore," (*Giornale degli Economisti e Rivista di Statistica 51*: 1-26) Homewood: Richard D. Irwin, pp. 27-56.

Sonnenschein, H. (1971), "Demand theory Without Transitive Preferences," in Chipman, J. S. and others (eds.), *Preferences, Utility and Demand*. New York: Harcourt Brace Jovanovich, pp. 215-23.

Tobin, J. (1958), "Liquidity Preference as Behavior toward Risk," *Review of Economic Studies 25*: 65-86.

Varian, H. (1983), "Nonparametric Tests of Models of Investor Behavior," *Journal of Financial and Quantitative Analysis 18*: 269-78.

von Neumann, J. and Morgenstern, O. (1947), *Theory of Games and Economic Behavior*. 2d ed. Princeton, Princeton University Press.

Wong, S. (1973), "The 'F-twist' and the Methodology of Paul Samuelson," *American Economic Review 63*: 312-26.

II. History:
The Role of Failure in
Early Scientific Development

5. FAILURE AND EXPERTISE IN THE ANCIENT CONCEPTION OF AN ART*

James Allen

I

The ancient notion of an art (τέχνη) embraced a wide range of pursuits from handicrafts such as shoemaking and weaving to more exalted disciplines not excluding philosophy (cf. Plato *Gorgias* 486b; Hippolytus *Refutatio* 570.8 Diels; S.E. M II.13). Nevertheless, there was enough agreement about what was expected of an art to make possible debates about whether different practices qualified as one. According to the conception which made these debates possible, an art is a body of knowledge concerning a distinct subject matter which enables the artist to achieve a particular type of beneficial result (cf. especially Heinimann 1961). Obviously, the failure of a practitioner to achieve the aim of an art can form the basis of a fairly simple challenge to his artistic competence. But when even the practitioners who most fully satisfy the internal standards of a profession are prone to failure, the artistic status of the practice itself can be called into question. Doubts then shift from the competence of an individual practitioner to the assumption that his practice is an art.

The aim of this paper is to sketch in very rough outline the history of the problem posed by failure. As we examine the responses it elicited from different schools of thought, it will become clear that failure was the common stimulus behind a number of important developments in the ancient conception of an art. The developments we will be concerned with are not the specific and substantive changes in the theory and practice of particular arts which failure may have prompted, for example, the kind of improvement in surgical practice we would expect surgical failures to suggest. Rather, our concern will be with the way in which the regular and apparently ineradicable occurrence of failure led to revisions in the conception of an art itself and the associated notions of artistic knowledge and expertise. In particular, I want to suggest that one such development was the emergence of a crude but recognizable notion of probabilistic knowledge.

II

In order to see how failure could form the basis of a challenge to the artistic status of a practice, we need to understand what was involved in the claim that an art is a body of knowledge. For it was possible to concede that a practice which occasionally fails to secure the result at which it aims is very useful, but still maintain that it cannot involve knowledge in the way required of an art if it is prone to failure. What the requirement that an art be a matter of knowledge amounted to will become clearer if we see how some of its implications were used to challenge the artistic status of certain practices. So, for example, it was agreed that because an art is a body of knowledge it is typically transmitted by teaching instead of being a natural endowment or the result of unsupervised practice. Thus those who professed to practice or teach an art could be challenged to show that the ability at issue was a result of instruction and did not come from another source. Socrates challenges Protagoras's claim that virtue is an art and thus teachable in just this way (Plato *Protagoras* 319a10 ff.). At first, Socrates argues that it is not teachable because we do not see it being taught or learned in situations where we would expect to see it passed on if it were teachable. Socrates's argument here corresponds to a standard move in debates over the artistic status of different practices: For example, the fact that some people are able to become good orators without professional instruction, while many who had received the training fail to become good speakers, was used to suggest that artistic instruction was unnecessary and ineffective (cf. Philodemus *Rhet.* II, 71; II, 97; Cicero *De orat.* I, 5; I 91: Quintilian *Inst. orat.* II, 17, 11; S.E. M II, 16). Instead, it was suggested, ability of this kind was not a matter of art, but of natural endowment. Protagoras's response, which became the orthodox answer to this challenge, was to point out that what is actually observed is not the absence or failure of teaching, but variations in ability not directly correlated with the amount or quality of the teaching received; and he argued that different levels of accomplishment in the arts may reflect different levels of natural aptitude, even though the ability in question is a matter of artistic knowledge, because a student's ability to benefit from teaching is subject to natural variation. Defenders of the arts could also safely concede that practice and hands-on experience had an important part to play in the successful practice of an art (cf. Aristotle EN 1103b8-13, 1179b20-21).

The orthodox account assigned a part to each of these three factors; natural ability, practical experience, and knowledge (cf. Isocrates *Adv. soph.* 13, 17; *Antid.* 15, 187-192; Plato *Phaedr.* 269d.,72a; cf. Philodemus *Rhet.* I, 51 col. 25). It enabled the teacher of an art to (a) answer the charges which might arise from the poor performance of some of his students by blaming their lack of natural aptitude or diligence and (b) allow that some prospective artists may succeed with very little, perhaps in exceptional cases none, of the kind of instruction he offers, all without diminishing the value of the knowledge he promises to teach. For, though it concedes a substantial role to factors other than knowledge and teaching, this account continues to make the role of knowledge primary: If it is correct, natural aptitude enables the aspiring artist to acquire the knowledge that makes up an art, while practice develops his ability to exercise that knowledge to the best possible effect.

But those who laid claim to artistic knowledge were faced with another, more difficult challenge. Socrates finds it easier to accept the account just given and concede that virtue is an art and consequently teachable than to agree that what Protagoras has to offer qualifies as an art (*Prot.* 361a-c). This is because there was much scope for disagreement about the standards a set of teachings had to meet in order to qualify as a complete body of artistic knowledge: Depending on the view of knowledge to which appeal is made, different sets of standards, some of them quite stringent, could be imposed on practices for which the status of an art was claimed. In the *Gorgias*, Socrates puts forward certain arguments that rhetoric is not an art which were to be repeated for centuries thereafter. Especially important for our purposes is the argument that rhetoric is not an art because, instead of being a matter of knowledge as a true art must be, it is a matter or mere experience and habit. An art, Socrates insists, must give a rational account of the real nature of its subject matter, an account which enables the artist who has mastered it to specify the underlying causes in light of which he acts as he does (*Gorgias* 465a,501a; cf. *Phaedr.* 270b). The sharp line between experience and knowledge drawn here was an innovation. Another participant in the dialogue, Polus, explicitly held that artistic knowledge was a form of experience (*Gorgias* 448c,462b). The line was not always drawn so sharply afterwards either. Though he did not identify art and experience with each other, Aristotle insisted on a very close relation between them (cf. *An. post.* B 19, 100a5; *Met.* A 1, 981a4; *Rhet.* 1359b30-32). And as we know, the challenge laid down in the *Gorgias* was later taken up by the medical Empiricists, who set out to show that unassisted experience could give rise to the art of medicine (an approach which could be and was applied to other arts). Such efforts were justified by the very reasonable countercharge that Plato's positive conception of artistic knowledge, at least as it was reflected in some of his dialogues, imposed standards which were impossibly high, if not completely inappropriate. The model of the measuring art held up in the *Protagoras*, for example, seems inappropriate for practical wisdom and arts of the kind to which it was compared (356a ff.; cf. Aristotle EN 1094a11-16; 1098a26; 1104a2); and it could be argued that the emphasis on the mathematization of the arts in the *Philebus* involves a similar appeal to inappropriate standards (56a-57e).

Nevertheless, it needs to be emphasized that Plato's objections to rhetoric had a basis in the common conception of an art. Roughly speaking, the common conception suggests a view something like the following: While someone can pick up a knack on the basis of experience that enables him to bring about a certain kind of desirable result more frequently than a layperson who lacks that experience, if the ability depends on nothing more than a few rules of thumb, a few memorized procedures adapted to

stereotypical situations, it does not yet amount to an art. The same view
is behind Aristotle's comparison of teachers of eristic and rhetoric to people
who promise to teach the knowledge of how to prevent feet from suffering,
but give their students a pile of shoes instead of the art of shoemaking: by
giving their students speeches and arguments to memorize, these teachers
give them the products of art, not the art itself (*Soph. el.* 183b36-184a8).
A body of artistic knowledge must be, somehow, systematic and complete
(cf. *Resp.* 342d10, 345d; 360e-361a; *Phaedr.* 269d-e. 272a-b). It must give
the artist more than a few effective procedures; it must give him a systematic and organized mastery of the procedures that will secure the end of
his art, a requirement which was well expressed by Aristotle's demand that
the true artist have at his disposal all the possible means of securing the
end at which his art aims so that he need never omit any procedure that
may contribute to success (*Rhet.* 1355b10-11; *Top.* 101b5-10). And this
requires not only that the artist learn all that he is taught—no doubt some
of Gorgias's students accomplished this—but that the art transmitted to
him embrace all possible means of success. And this in turn may call for a
theory in which the means at the disposal of the artist are connected with the
nature of the matters with which he must deal, so that the requirements
imposed by Socrates in the *Gorgias* may not be out of order after all.

Thus defenders of a practice whose artistic status had been challenged
needed to show that the body of knowledge on which they relied, in addition
to being of some use, was complete and systematic according to the appropriate standards. The kind of knowledge Plato sometimes demands of an
art promises to satisfy this requirement. Clearly, unsystematic experience
will not. What remained controversial was whether experience could be
systematized and organized so that the requirement can be satisfied.
Consequently, disputes about the artistic status of practices like rhetoric
and medicine frequently became controversies about how the requirement
that an art involve a complete and systematic body of knowledge should
be understood. This fact has an important consequence which we must
bear in mind when we turn to the problem posed by failure. As Aristotle
notes, the requirement of completeness involves a certain amount of idealization (*Top.* 149b24). Discussion was not so much about the arts in their
present state, where there might be significant room for improvement, as
in an envisaged ideal condition of completeness or perfection. Thus responses to the problem of failure had to show that its occurrence is compatible with possession of an art in its ideal form (cf. Cicero *De orat.* I, 76).

The kind of failure at issue occurs when an artist does not achieve the
end at which his art aims. Examples include the failure of a navigator to
pilot his ship safely to its destination, of an orator to persuade an audience
or of a physician to cure a patient. This is not quite precise enough,
however. The Hippocratic writers often point out that the medical art is not
all-powerful and cannot be expected to cure all ailments or restore every

patient to health; some patients in some conditions are beyond the power of medicine, and the Hippocratic physician is urged to distinguish carefully between curable and incurable patients (*De arte* c. 3; *Prognosis* c. 1; *Vet. med.* c. 9). In general it was thought to be a characteristic of the artist to be able to distinguish between the possible and the impossible in his area of expertise (Plato *Resp.* II, 360e; Stobaeus IV, 38, 9 Wachsmuth). And so far as it goes, this seems absolutely right: The inability to achieve impossible results is not a failure; inevitably nature imposes limitations on the power of art. Rather, we are only really faced with failure, for example, when a physician's attempt to cure a patient does not succeed or a prognosis of his turns out to be false.

What is puzzling about the Hippocratic account, or rather the extremely crude sketch of it just provided, is the way in which it seems to envisage a physician who knows which patients he can, and which he cannot, cure: a physician who will at no point undertake cures with less than complete confidence of success or qualify his prognoses with reservations. (See *Vet. med.* c. 9 for a conspicuous and important exception.) There seems to be no room in this account for good faith efforts undertaken with a reasonable expectation of success which nevertheless fail. Understood in this way, the conception of an art would seem to require the artist to be invariably successful, unless a clearly extrinsic circumstance interferes. A practitioner's claim to be a genuine artist could then be called into question by a single failure. More significantly, the artistic status of practices like medicine and rhetoric, whose best qualified practitioners often fail, will also be questioned if this standard is in force: They do not, as it was put, deliver what they promised (cf. Quintilian, *Inst. orat.* II, 17, 22; Philodemus *Rhet.* I col. 5). Yet by characterizing rhetoric as the "manufacturer of persuasion," rhetoricians like Corax, Tisias and Gorgias seemed to invite this charge (*Gorgias* 463a; *Philebus* 58a; cf. Quintilian II, 17, 23). And some physicians seem to have made claims nearly as strong, or at least to have been faced with similar expectations (Hipp. *Acut.* c. 9; *De arte* c. 8). But the strong claims made for these arts, and the high expectations they faced, were not just a result of their practitioners' vanity. The artist's claim to a complete and systematic body of knowledge could be taken to imply immunity from failure, not omnipotence, of course, but the ability to foresee when his efforts would be successful and when not. If the artist really understood the nature of the matters with which he dealt, it could be argued, he would not raise hopes that were destined to be frustrated. As in many arts, failure would indicate a gap in the practitioner's knowledge that needed to be filled before he could count himself a complete artist. In many areas, mastery of the appropriate art did guarantee success, while failure indicated a gap in the practitioner's training. Why, it could be asked, not in medicine or rhetoric as well?

The way in which Thrasymachus escapes from a difficulty Socrates puts in the way of his view that justice is "the benefit of the stronger" makes it clear just how strong the connection between artistic expertise and success could be taken to be (*Resp.* 340d-341a; cf. 342b). Socrates raises a problem for Thrasymachus's view by pointing out that the strong sometimes err in deciding what will be to their benefit, so that if compliance with their wishes is sufficient for justice, as Thrasymachus also seems to think, the same act may turn out to be both just and unjust. Thrasymachus escapes this difficulty by claiming that a man is not properly speaking stronger when he makes an error about what is to his advantage, just as a physician is not properly speaking a physician when he misses the mark (ἁ μαρτάνει). Thus Thrasymachus is able to appeal to a view, which he expects Socrates and the other participants in the dialogue to share, according to which an art cannot be exercised without leading to success: Failure must imply the absence, if only temporary or partial, of artistic knowledge. And Socrates appears to adopt a similar view himself in the *Euthydemus* when he asserts that wisdom of the kind that practitioners of different arts and skills draw on makes men do well always so that they never err in act or result (τυγχάνειν) (280a-b).

III

Nevertheless, at a fairly early stage it was acknowledged that failure is a legitimate feature of certain arts in a way that it is not of others. We are permitted to conclude more or less automatically, in the absence of appropriately unusual circumstances, that someone who fails in an attempt to make a pair of shoes is not a master of the art of shoemaking. On the other hand, the occasional failures of a navigator, a physician or an orator to achieve a desired result need not show that these practitioners are not artists or that the knowledge of which they make use does not qualify as an art. Arts in which failure was permitted to occur in this way came to be called stochastic from the verb meaning to aim: στοχάζεσθαι(Alex. Aphr. *In top.* 32, 12 ff.; Quaest. II 16; Philodemus *Rhet.* I 26, 6 ff.; 59, 17 ff.; 170, 3 ff.; S.E. M II 13). Two closely related applications of the word were involved. The expression was used to indicate that the relation between these arts and their ends was one of aiming, not always of hitting ([Galen] *Medicus* XIV, 685 K; Alex. Aphr., *Quaest.* II 16). But "*stochasmos*" was also used to mean conjecture, the kind of educated guesswork an artist employs in the absence of conclusive evidence (Hipp. *Vet. med.* c. 9; Plato *Gorg.* 463a. 465a; *Phlb.* 55a-56b). The underlying motivation was the same of course, for the success aimed at, but not always achieved, in conjectural reasoning is the truth of the matter at issue.

Here, then, is the first development in the conception of an art prompted by failure. In certain arts success was downgraded from a necessary condition in the assessment of the artistic status of a practice and its practitioners. One way of doing this involved a distinction between the end (*telos*) and the function (*ergon*) of an art (cf. Alex. Aphr., *In top.* 32,12-34,5; Quintilian II

17,22-25). In the nonstochastic arts, end and function are coextensive; the function cannot be fulfilled without the end also being achieved. In the stochastic arts, however, it is possible for the artist to fulfill the demands of his art without achieving its end. Thus the end of the medical art is to save the patient, while its function is to do everything possible to save the patient. An artist must always be successful *qua* artist by fulfilling his function, but a stochastic artist may do so without always achieving the end of his art. This insight may already be implicit in the distinction Socrates draws in his conversation with Thrasymachus between two types of benefit (συμφέρον) produced by the arts. These are the benefit of the object of the art's concern, e.g., the health of the patient in the case of medicine and the benefit of the art itself. The latter is achieved as long as the art is complete and in no respect deficient (*Resp.* 341d6-11; 345d1-5). What needs to be added is an explicit recognition that an art can hold on to its own benefit even when it does not achieve the first type of benefit and an explanation of how it can meet internal standards of completeness which are not invariably linked to the achievement of that external benefit.

This innovation in the conception of an art made it possible to concede artistic status to the stochastic practices without sacrificing the part of that conception to which Thrasymachus appeals. The true artist never misses his mark (ἁ μαρτάνει), but the mark he always hits is the perfect satisfaction of the demands of his art, not the beneficial result to the achievement of which his art is dedicated. The point can be made in Thrasymachus's own language.[1] In support of his point that a person who fails is to that extent not an artist, he maintains that an artist, insofar as he is an artist, never fails (*Resp.* 340d7-e1; cf. 342b5-6). The stochastic artist agrees. However, he understands the point of this remark not to be that an artist properly exercising his art never fails without qualification, but that he never fails *as* an artist, *qua* artist or at his art.

More is called for if the stochastic artist is to defend his claim to the title of artist without qualification, however. If standards are raised so that a practice qualifies as an art only if it involves the grasp and application of a body of knowledge that guarantees success, the stochastic arts will still not qualify as arts in the truest and strictest sense of the term. And critics of the stochastic practices were sometimes willing to grant that there was a weaker and more extended sense of the term according to which the stochastic practices qualified as arts, as long as it was understood that they were inferior to arts properly so called (Plato *Phlb.* 55e-56a; Philodemus *Rhet.* I 72, col 40; cf. I 59, 19 ff.; I 70, 15 ff.). If the defense of the stochastic arts stops here, however, the stochastic artist will have won a certain measure of safety for his claim to artistic competence at the cost of accepting an inferior status for his art. Our concern is with artists and philosophers who did not fall in with this suggestion. What was needed, and what they attempted to supply, was an account of artistic knowledge compatible with artistic failure. If such an account was to succeed, it needed to face squarely the motivations behind the tendency to link artistic knowledge with success. To do so, it had to give convincing answers to the following two connected sets of questions:

1) What accounts for the fact that complete bodies of knowledge are not sufficient for success in the stochastic arts in the way comparably complete bodies of artistic knowledge are in the nonstochastic arts? In particular, why is the imperfect record of success in the stochastic arts not to be attributed to the incomplete character of the knowledge they rely on, to gaps in that knowledge which have to be filled to produce a complete body of knowledge and a genuine art?

2) What is it that the stochastic artist *knows* when he knows that the measures he has adopted are the right ones if knowing it does not involve knowing that they will succeed? In other words, how is the stochastic character of an art reflected in the form of the knowledge on which it relies?

Though answers to the first question came in cruder and more sophisticated forms, they tended to agree that the imperfect record of success in the stochastic arts was due to a special feature of matters they dealt with. Unlike the processes with which the nonstochastic arts are concerned, those the stochastic artist deals with are by nature variable and lacking in fixity. Since this variability escapes precise formulation in the precepts of an art, the application of the precepts of a stochastic art is a much less straightforward business than is the case in the nonstochastic arts (cf. Dionysius Hal. *De comp.* = *Art. script.* B, VII, 23 Radermacher). Thus, if they were to succeed at their art, stochastic artists needed to do more than acquire a mastery of the formal precepts of their art; they also needed to develop a sensitivity to the peculiar features of particular situations, a sense of the opportune moment (ὁ καιρόσ) which enabled them to undertake the right procedures, at the right time in the right way (cf. *Phaedr.* 272a). Since this ability cannot be incorporated in the formal precepts of an art, it has to be built up by practice and hands-on experience (Isoc. *Antid.* 15, 184; cf. Aristotle EN 1104 a 3-10). This point was frequently made by contrasting the stochastic arts with the art of writing the letters of the alphabet, an example of the kind of art whose subject matter is such that the mechanical application of the art's rules guaranteed success (Hipp. *Vet. med.* c. 20; *Loc. hom.* c. 40-45; Isoc. *Adv. soph.* 13, 10; 13, 12; Aristotle EN 1112b1-6; Phld. *Rhet.* I, 70, 30 ff.). What is more, because of this variability in the nature of the matters they deal with, the stochastic arts rely on precepts which are characterized by a certain roughness and imprecision; they allow for exceptions. But there was a strong tendency to treat the element of variability as a relatively small one. This explains the appeals to the for the most part (ὡϲ ἐπὶ τὸ πολύ, κατὰ τὸ πλεῖϲτον) frequently made in accounts of the stochastic arts. The precepts on which the stochastic artist relied were supposed to hold for the most part (cf. Aristotle *Met.* 1027a20 ff.; Alex. Aphr. *In an. pr.* 39, 25). And it was frequently also assumed that the true stochastic artist, though permitted a few failures, must succeed for the most part (cf. Isoc. *Antid.* 15, 184; Phld. *Rhet.* I, 58, 7 ff.; II, 125 fr. 9; Diogenius apud Eusebium *Praep. evang.* IV, 3, 1; S.E. M II, 13; Galen *De meth. med.* X, 58, 17 K).

At least in its present form, this account is not a completely satisfactory response to the challenge the stochastic arts faced, however. It leaves several questions unanswered. In the first place, we will want to know what justified the requirement that the stochastic artist succeed for the most part. If nature places constraints on the amount of success that can be achieved in different areas, why should these constraints not vary from field to field (cf. Phld. *Rhet.* I 26, 13 ff.)? Moreover, we will also want to know why the constraints imposed by nature on the stochastic artist's rate of success, and on the level of precision his knowledge can attain, are ineradicable even from the art in its ideal form. In other words, why is it that claims about the lack of fixity characterizing an art's subject matter do not just mean that the knowledge it requires is harder to obtain, or that the knowledge on which it relies in its present form is still incomplete? Finally, we will want to know why, even if it is conceded that nature permanently bars some arts from shedding their stochastic character, the roughness and inexactitude which is a permanent feature of the precepts they rely on does not mean that they are not really arts in the truest and strictest sense of the term (cf. Cicero *De orat.* I, 92). Defenders of the stochastic arts need to give us a reason to concede that the stochastic arts involve artistic knowledge, albeit knowledge of a special kind, instead of concluding that the precepts on which they rely, though not entirely value-less, do not amount to a body of knowledge. In other words, we need to see how the introduction of a type of knowledge with special stochastic characteristics can be justified.

One ancient account of the stochastic arts along the lines sketched above handled these questions very well, that of Aristotle and his followers. As we have already seen, Aristotle is to be credited with a clear statement of the crucial insight that, in the absence of procedures which guarantee success, it is still possible to develop a complete theory of all the means at the disposal of the artist. The result was that the practitioner who had mastered all of them, and developed the ability to select the right ones for the right occasions, was as entitled as a nonstochastic artist to lay claim to a complete body of artistic knowledge: Though he might fail to achieve a particular sought-for result, he will never fail at his art (*Top.* 101b5-10; *Rhet.* 1355b-10-11, b25-6). And Aristotle's account of the stochastic arts also crucially relied on for the most part precepts (*Met.* 1027a20; cf. Alex. aphr. *In an. pr.* 39, 16 ff.).

But the best way to see how the Aristotelian approach resolved the difficulties which faced the stochastic arts is to turn to the account found in Alexander of Aphrodisias, who, without departing in any essential point from ideas already present in Aristotle, gave these issues more explicit attention than Aristotle had. Alexander distinguished the stochastic arts from the nonstochastic arts as others had distinguished them from fixed or firm arts (or *epistēmai hestekuiai* or *pagioi*) (*In top.* 34, 2; cf. Philodemus *Rhet.* I 26, 6 ff.; 59, 17 ff.; 170, 31 ff.; S.E. M II 13). In the nonstochastic arts the procedures through which, and only through which, their aims are achieved are themselves determined: If the procedures are carried out correctly, the end aimed at must result, and its occurrence is a sure sign that they have been carried out correctly (*In top.* 33, 12-15). On the other hand, the procedures through which the ends of the stochastic arts are realized are not determined in the same way (*ibid.* 33, 18-20). If the end of the art is to be achieved, the cooperation of circumstances which are beyond the power of the art to control or predict are needed as well (*ibid.* 34, 1; *Quaest.* II 16); and the end may sometimes come about due to luck, when the art is not exercised. For this reason, the stochastic artist must act under conditions of uncertainty when he attempts to achieve his end or hazards a prediction. This is why it is worthwhile for him to do all that is possible, because it cannot be known that the measures decided on will not be effective.

So far, this account is very much like the standard treatment of the stochastic arts outlined above. What makes it different is the support it receives from the Aristotelian conception of nature. According to that view, knowledge of certain parts of nature can only take the form of for the most part truths, even at the level of first principles. And because these principles turn out to mirror the nature of the matters they describe, when organized and systematized in the appropriate way, they can qualify as a body of genuine knowledge, the kind of knowledge Socrates demanded of an art. Thus the appeal to the for the most part in the Aristotelian account does not look, as it sometimes does in other accounts, like an ad hoc attempt at damage control.

The Aristotelian view of nature is familiar enough in its broad outlines. A natural substance—the clearest examples are provided by individual living things—is the kind of thing it is because of the possession by its matter of the appropriate kind of form. This form is its nature and is responsible for the possession by the thing of the capacities characteristic of its kind. The view is a strongly teleological one: The capacities are present in a thing because they enable it to fulfill its function which, in the case of living things, typically consists in the active exercise of the more important and characteristic of these capacities. Moreover, natural processes are typically directed at, and take place for the sake of, the realization of forms of different kinds in matter and at the exercise of the capacities for which, once realized, these forms are responsible. This is so even when that realization does not take place. There are natural processes which take place for the sake of ends which are not achieved. Nevertheless—and this is the most important point for our purposes—natural processes reach the ends at which they are directed for the most part. Thus genuine scientific knowledge, knowledge of what things are and do by nature, only embraces what things do always or for the most part; only these things belong to their nature and can be explained by reference to it. Aristotle is content to leave exceptions to the for the most part patterns enforced by the nature of things as anomalies. This is not to say that they must remain unexplained. At least typically, exceptions to patterns which hold for the most part are due to outside interference in the normal and natural course of events; and often enough it will be possible to say what the source of interference is and why it had the effect it did.[2] But such explanations will always take the form of stories about how the normal and natural course of events was hindered. Aristotle does not envisage, even as an ideal, a scientific understanding of the world which puts all outcomes on an equal footing. According to such a view, there would be no exceptions. All explanations in which an object or event is held responsible for interfering with the normal outcome could, in principle, always be replaced with another according to which the actual outcome is the normal and perhaps even required outcome of the process interfered with and the interfering circumstances. But this is not the path Aristotle took.

I have emphasized the teleological character of the Aristotelian approach because, were it not for his teleological commitments, Aristotle's failure to adopt such an ideal might seem to involve an *arbitrary* rejection of a conception of nature which goes along with it. According to this conception, a thing's nature governs all of its behavior in the circumstances in which it finds itself. The view could come in more and less deterministic versions. According to the first, a thing's nature is a set of dispositions which in combination with the thing's circumstances determines how it will behave. According to the second, a thing's nature is something like a collection of propensities which make different types of behavior more or less likely in a given set of circumstances. Neither version of this conception of nature lends much support to Aristotle's view that what occurs by nature does so always or for the most part only. And the corresponding restriction of knowledge to what happens always or for the most part seems equally hard to justify (*Met.* 1027a20 ff.; cf. Alex. Aphr. *In an. pr.* 39, 16 ff.). For the way a thing behaves for the most part would be due to its nature no more than the way it behaves, e.g., half of the time in a given set of circumstances. And a full account of its nature would explain the behavior it exhibited half the time or infrequently just as much as the behavior it exhibited for the most part. Yet in the teleological context in which it developed, the Aristotelian approach makes a great deal of sense. If we think of the nature of a thing as a set of capacities adapted to a specific set of functions, the privileged status accorded to the events which are part of the realization of a thing's nature, or conditions or concomitants of its realization, is a reflection of the fact that these events belong to its nature in a way that others do not. And this account shows how knowledge of the for the most part can be knowledge of the nature of things and hence real knowledge. Truths about what happens for the most part can be recast as truths about the nature of things with the proviso that events come about in accordance with nature only most of the time. Outcomes which do not occur in connection with the natures of things in the way for the most part occurrences do are not more recondite realizations of nature, but interruptions in the normal and natural course of events (or undetermined by the nature of the thing); and their explanation will always remain somewhat haphazard and unsystematic in comparison with the explanation of what accords with a thing's nature.

It should now be clear how the Aristotelian view of nature made a convincing defense of the stochastic arts possible. Stochastic artists set in motion processes which only work in the desired way most of the time, and this is because of the nature of the matters involved. This account leaves room not only for failures which are not the result of artistic error, it makes it possible to see how the stochastic artist can be the master of a body of knowledge no less secure and certain, and no less complete, than the knowledge at the disposal of the nonstochastic artist and still be entitled to entertain hopes which will be frustrated and hazard predictions will not come true. For the nature of the matters which are his concern is not always fully realized. But claims about what accords with their different natures are no less true for that reason. Knowledge of the nature of things, in his area, only translates into knowledge of what happens for the most part in actual fact. And this conception of nature also dictates the answer to the second question posed above: The stochastic character of the art is reflected in the for the most part provisoes attached to the account of the nature of the matters it deals with (cf. Alex. Aphr. *In an. pr.* 39, 25).

At the beginning of this paper I suggested that the ancient debate about the stochastic arts was a good place to look for the emergence of a notion of probabilistic knowledge. Though it might at first seem that the Aristotelian approach would be hospitable to such a development, we can now see that the Aristotelian view of nature stood in the way. According to that view, events in nature are classified under two heads: Either they are part of the course of events normal relative to the realization of some nature, or they are exceptions to what is normal and natural relative to that nature. Thus the Aristotelian account provides no motive for distinguishing a *range* of different relative frequencies and using the knowledge gained in this way to assess the likelihood of events of particular types.

In fact, both variants of the other conception of nature sketched above appear more hospitable to the development of probabilistic views. Of course, if the first, more deterministic version of the alternate view is adopted, scientific knowledge will ideally consist of a set of exceptionless generalizations which, when applied to a sufficiently complete and detailed description of a state of affairs, would make it possible to see subsequent events as the required outcome of that state of affairs. But if there are limitations on human knowledge, so that we must conjecture on the basis of partial information, this approach is compatible with an epistemic view of probability which, while assuming that there are hidden factors at work that determine observed outcomes, makes those outcomes more or less likely on the basis of observed relative frequencies of occurrence. If the second, less deterministic version is adopted, it is even easier to see how a probabilistic account of knowledge might be suggested, since scientific knowledge at the ideal limit would be of natural propensities that could not be eliminated from any description of the natural world no matter how fine-grained. In neither case would we expect the concentration on outcomes occurring always or for the most part found in the Aristotelian account.

IV

Although the Aristotelian view of nature stood in the way of a probabilistic view of artistic knowledge, it provided a very effective justification for the privileged status granted to the for the most part in the traditional account of the stochastic arts. Without the underpinning provided by the Aristotelian account, the reliance of the traditional account on the for the most part became harder to justify, however. And it did not go unchallenged. The following objection is reported by Philodemus: "[T]he physician who saves ten out of a hundred difficult cases, e.g., of elephantiasis, is not successful for the most part, but he is a good and worthy physician" (*Rhet.* II 120). The point is a good one, but the appeal to for the most part success in accounts of the stochastic arts proved remarkably persistent and was still being criticized centuries later by Galen, who complains: "[If] someone errs once in twenty times he has fallen short of invariable certainty, but he is better than the man who errs twice as that one is better than the man who errs three times...and yet all have adhered to the expression 'for the most part'" (*In Hipp. prog.* CMG V 4,2 204, 19 ff.; cf. 377, 14 ff.; *De meth. med.* X, 58, 17 K).

These objections call attention to two different defects in the for the most part view. That view was motivated in part by the need for a reliable way of distinguishing the (stochastic) artist from the layperson. The idea seems to have been that though the layperson may chance on an occasional success by proceeding unsystematically, by and large he will fail, while the opposite is true of the trained artist. However, as Galen's objection makes clear, this is not sufficient. An adequate account of artistic expertise must distinguish the artist not only from the layperson but the semiskilled practitioner and the novice as well, as vague references to success for the most part cannot. This is part of the requirement that an art must be somehow complete. On the other hand, as the objection cited from Philodemus shows, success for the most part cannot be a necessary condition for artistic mastery either. Insistence on success for the most part does not allow for the fact that nature may impose limits on an art which prevent it from achieving success for the most part without rendering its development pointless. Aristotle's appeal to the for the most part was a consequence of his conception of nature; it reflected a condition imposed on the arts by nature. If an account of the stochastic arts invokes a different conception of nature, it can dispense with Aristotle's emphasis on the for the most part, but something corresponding to his insistence that the artist must have achieved a complete mastery of his method to the point of being able to accomplish everything nature permits to achieve his aim is essential. And another view preserved by Philodemus takes account of this fact (*Rhet.* I 25, 32 ff.): "[T]he stochastic artist achieves his aim as much as nature permits, not [necessarily] for the most part or mostly but much more than the layperson." And this is possible for an artist only when he has fully mastered an art which is itself complete.

V

Of course, agreement on this point still left much room for disagreement about what were the standards a body of knowledge must satisfy to be accounted complete. The positions different schools took on this question reflected their views on general epistemological issues. Those who believed rational insight into the underlying nature of things was obtainable insisted that artists will only be able to claim that they are doing all that nature permits when they have a theory which makes it clear what constraints nature imposes on artistic practice. On the other hand, those who were dubious about the possibility of attaining such knowledge, and about its practical benefits even if it were obtainable, opposed this conception of completeness and championed an alternative empirical account. Though of interest to practitioners of the other arts as well, the issues in dispute between these two points of view were most thoroughly explored in ancient medicine, where a vigorous debate sprang up between schools of medical Rationalists and Empiricists. For this reason, the remainder of this paper will consider how the medical Rationalists and Empiricists used their theories to reconcile artistic knowledge with failure.

One reason to look closely at this dispute is that both schools are more likely to have been hospitable to a probabilistic account than Aristotle and his followers. Both were in a position to take account of the criticisms of the for the most part, which blocks the development of such a view in the Aristotelian framework. And the advantages of a probabilistic account when it came to reconciling artistic knowledge and failure should be obvious. The appeal to different relative frequencies in the passages cited from Galen and Philodemus do not yet amount to such a view, however. The relative frequencies cited there were used to evaluate artistic competence and made a kind of external probability judgement possible, *external* because they concern the exercise of artistic knowledge as a whole without reference to the character of the knowledge itself or the inferences it supports. The fact that an observer might expect success from an artist in proportion to the relative frequency of his past successes is compatible with an artistic method which leaves little or no room for degrees of expectation on the part of the artists themselves. It would still be possible to opt for a Thrasymachean account and make human error responsible for failure. Probabilistic judgements will then concern the likelihood of human error, and not the relative frequency of different outcomes in the processes with which the stochastic arts are concerned. Of course this was not a very attractive option; nor was it a way of *defending* the stochastic arts. Thus the acknowledgement that a certain rate of failure is to be expected in different artistic practices put a certain amount of pressure on the stochastic arts to internalize frequency based judgements, a strategy which was adopted by the medical Empiricists, who made relative frequencies the object of artistic knowledge. Yet internalization was not inevitable. The medical Rationalists often chose, if you will, to insulate knowledge from failure in practice. And it is to their views that we will first turn before proceeding to consider the Empiricists' criticisms of these views and their own positive account of artistic knowledge.

The Rationalists claimed that their theories gave the physician knowledge of the underlying causes of disease. And they went on to argue that this knowledge would enable the physician to select the remedies and treatments naturally adapted to counteract the forces producing and maintaining the diseased condition. Thus according to their view, the method of medicine was a matter of making inferences from the manifest symptoms of the patient to their hidden, underlying causes and using the information gained in this way to infer the appropriate treatment (cf. Galen *De sect. ingred.*, SM III, 7; *De causis continentibus* = 141, 1-3 Deichgräber).

Rationalist thinkers developed their theories in a philosophical climate strongly influenced by the Stoics. In particular, the Stoics' insistence on a deterministic account of nature raised the stakes for those who wished to lay claim to scientific knowledge of natural processes. According to the Stoa, every event, no matter how apparently insignificant and fortuitous, was the completely determined and rationally required outcome of preceding events (cf. Cicero *Div.* I, 127). And it could be argued that the conception of knowledge which corresponds to this view, unlike the Aristotelian view it supersedes, leaves no room for unfulfilled predictions and failed artistic procedures. Though there might be matters beyond the reach of the physician's knowledge or the powers of his technique, it was hard to see how he was entitled to failures in prognosis and therapy in those matters where he did claim rationally warranted knowledge. When failures occur in this kind of situation, the choice seems to be between accepting the failure as a refutation of the theory or protecting the theory at the expense of finding fault with the physician. Thus the problem posed by failure was particularly acute for the Rationalists, who laid claim to knowledge of the real underlying nature of sickness and health.

Their response was to adopt what I have called a strategy of insulation. Though willing to admit that medicine was a stochastic art, they refused to concede this was due to an element of uncertainty in medical theory. They maintained that the theorems which make up medical knowledge were certain and universal; it was practice which introduces a stochastic element into medicine ([Galen] *De opt. sect.* I, 114 K). In line with this view, Erasistratus, e.g., distinguished the epistemic or scientific parts of medicine like physiology and aetiology from stochastic parts like diagnosis and therapy ([Galen] *Medicus* XIV, 684 K). This move is an attempt to defend medical theory by making the application of theory, and not theory itself, responsible for failure. If the explanation is successful, an important class of failures will not reflect discredit on medical theory.

The insulating strategy was not an ad hoc move to defend the Rationalists' claim to certain knowledge by refusing to submit their theories to the test. By their own lights, the Rationalists had the best possible reasons for absolving medical theory of responsibility for the stochastic character of medicine. Rational theories, they maintained, are inevitably abstract and general in character. A rough fit between theory and the particular cases to which it is applied is consequently natural and inevitable. And medical views on the subject were backed up by philosophical views that tied knowledge to the universal and insisted on the corresponding ineffability of the individual (cf. Deichgräber [1956] 1984).

Rationalists and Empiricists agreed that the doctor must distinguish as many different kinds of patients and ailments as possible in order to distinguish corresponding kinds of treatment adapted to the special features of individual cases (Galen *De meth. med.* X, 207, 5 ff. K, cf. 209, 14 ff.). But the requirement that knowledge be restricted to the general and universal imposed a limit beyond which division could not proceed. In particular, a substantial residue of medically relevant features peculiar to the individual patient could not be the object of medical knowledge. According to the version of this position favored by Galen, the bodily nature of individual human beings was primarily a matter of their unique blend of humors and elements, or idiosyncrasy. Elsewhere idiosyncrasies were thought to account for a great many things about an individual human being, e.g., differences in perception, the amazing ability of some individuals to consume poisons which would kill most others and the like (cf. S.E. PH I, 181-3). Rationalist physicians appealed to them in order to explain the stochastic character of medicine. Medicine, Galen informs us (*De meth. med.* X, 209; cf. 206 K; *De dign. puls.* VIII, 774, 8-9 K; *De diebus decret.* IX, 932, ff.):

...has aimed conjecturally (*estochastai*) at the nature of the sufferer, and many doctors, I believe, call this idiosyncrasy and they all agree it cannot be grasped (*akatalepton*). For this reason they concede true medicine (ἡ ὀ'ντωδιατρική) to Apollo and Asclepius.[3]

The medicine of the gods envisaged in this passage would be able to take account of individual bodily natures in the way no human science could. The idea on which Galen's claim relies, that the individual in its complete concrete particularity eludes scientific knowledge, is an old one, which took different forms in different contexts. There is, e.g., a well attested Hippocratic tradition emphasizing the uniqueness of the individual patient and the need for specialized individual treatment (cf. Edelstein [1931] 1967). And, though the matter is of course complicated, there is a way of reading Plato and Aristotle according to which genuine knowledge is of the universal, of genera and species, but not of the individuals which fall under them (cf. *Rhet.* A 2 1356a31-33). Galen's version of Rationalism seems to represent a fusion of this kind of view with the very different outlook championed by the Stoa. Roughly speaking, a Stoic component in his view suggests that there is nothing imperfect or irregular about the nature of the individual or the particular natural processes in which it participates; they are completely determinate and rationally explicable, at least to divine reason. The other component, the restriction of knowledge to the universal and general, is turned into a limitation on human knowledge, not a claim about the imperfection and *consequent* unknowability of the natural world. Though in some ways similar to the Aristotelian view we have already met with, the resulting view is different in certain crucial respects. In particular, the Aristotelian view needs to appeal to nothing besides the nature of the matter in question to explain the imperfect efficacy of the stochastic arts; it is not necessary to distinguish human and divine knowledge as Galen does.

When this philosophical outlook is applied to medicine, a view something like the following results. Diseases come in genera and species. The generic nature of a disease clearly indicates its generic cure, e.g., if its nature is constrictive, dilation is indicated and so on. Thus the Rationalist is willing to speak of generic remedies (*De partibus artis medicativae* c. 5, p. 123, 10 ff., Lyons; *De meth. med.* X, 128 passim K). This indication is a matter of certain knowledge. But it is only of limited use, for all sorts of other factors need to be taken into account if the indicated treatment is to be adapted to the particular conditions at hand, e.g., the part of the body affected, the strength of the affection and so on. These can be regarded as differentia which determine the species of the generic affection. And further circumstances need to be taken into account, e.g., the age of the patient, the seasons and so on. So the Rationalists distinguish the principal indication of the treatment derived from the imbalance of the humors and subordinate indications from age and season (cf. Galen *De sanit. tuen.* CMG V 4.2 159, 17 ff. = 143, 21-30 Deichgräber; *Quod optimus medicus sit quoque philosophus*, SM II, 17 ff.; 6, 11-12). Thus a somewhat idealized picture of rational indication would include a tree with branches for the different species of the disease and sub-branches for other relevant differentiating features (cf. Galen *De meth. med. ad Glauc.* XI, 3, 12 ff.K). This tree would be mirrored by a tree which distinguished treatments adapted to the different varieties of the disease. Indication works at any level by proceeding from node to corresponding node. And viewed in this way, the indications can be seen to be rationally sanctioned and completely certain. The problem is that the lowest nodes on the tree of treatments do not determine the final choice of the precise treatment. There are no branches on the opposite tree at the level of individual patients and their idiosyncrasy. The individual's own nature and its interaction with all the factors in question is beyond the reach of the knowledge represented by the tree. Consequently the final determination is not a matter of rational indication but of conjecture. And the situation is made more complex by the fact that sometimes it is not only the final selection of the therapy which is conjectural; many diseases are also such that exact diagnosis is impossible. In a case where a patient is suffering from one of two diseases compatible with the manifest symptoms he displays, diagnosis will be a matter of conjecture as well. Thus stochastic conjecture enters into medical practice at two points: The final selection of the precise therapy adapted to the particular case at hand can be conjectural (*De meth. med.* X, 181, 17 ff., 206 K), and the diagnosis of the affliction is also sometimes conjectural (*De loc. aff.* VII, 581 K, *De plen.* VIII, 7 ff. K = 143 Deichgräber, *De sanit. tuen.* CMG 42 160, 24 ff. = 161 Deichgräber).

It should now be clear in outline how the Rationalist strategy of insulation worked. The fact that medicine was stochastic did not undercut the Rationalist's claim to certain knowledge of necessary connections in nature; rather, it showed that this knowledge was at a level of generality too high to guarantee perfect results in practice. Rationalists allow that they will occasionally fail in diagnosis and treatment, but insist that failures need not imply there is a gap or defect in their knowledge. Rational knowledge was perfectly firm and certain so far as it went; the problem was that it would only take physicians part of the way to the treatment they must administer. Rational indication works like an accurate but not particularly detailed map which provides clear directions to the general vicinity of the cure but not to its precise location.

But according to the Rationalists, the contribution of rational theory to practice does not end when conjecture takes over. An old view gives a considerable amount of credit to practitioners who have experience without theoretical knowledge because of the benefits that come from familiarity with particular cases; Aristotle held that they were more effective than those who were in the opposite situation and had genuine theoretical knowledge without experience (*Met.* 981a14). Galen too has much to say in praise of the Empiricists' close attention to particular cases and individual patients (*De meth. med.* X 62; 207, 5 ff. K = 143, 3 ff. Deichgräber). But he also held that physicians who had mastered rational theory had advantages in practice which were unavailable to the Empiricist, even when it came to the ineffable peculiarities of the individual patient. For even when it had ceased too provide explicit direction, rational theory continued to guide the practice of the physician. It is true that Rationalists must go beyond what theory can tell them and conjecture about the individual nature of the patient, but at least they know what kind of condition it is a particular form of which must be conjectured. For example, in cases involving an imbalance of the humors, the Rationalist's understanding of humoral pathology gives him an advantage when it comes to the conjectural part of his procedure. He will be able to take account of the factors which interact with the ailment and the idiosyncrasy of the patient much better than the Empiricist, who must rely on what has been previously attested in experience. For experience, though of particular cases, is always of particulars observed in the past and it offers no guidance in new cases in the respects in which they depart from previous cases.

Thus the Rationalists charged that the Empiricist will be at a comparative disadvantage because he had to rely on cumbersome syndromes into which all the possibly relevant factors have been added; he will be dependent on so called exact syndromes, whose recommended treatment is always the same. But these will be of no help to the physician in those situations, frequent enough to be sure, where the observable symptoms are compatible with several different conditions which may require different treatments or where a new feature of the affliction not previously encountered is present (*De loc. aff.* VIII, 14, 7 ff. K = 85 Deichgräber, *De sanit. tuen.* CMG 4 2 161 = 161, 26 ff. Deichgräber). The Rationalists charged that, without the guidance of theory, the Empiricist will be unable to make soundly based conjectures. On the other hand, the Rationalist for his part will use what he calls artistic conjecture, a method which falls somewhere between exact knowledge and complete ignorance but is guided by the former (Galen *De loc. aff.* VIII, 14, 7 ff. K = 143, 17 ff. Deichgräber). If correct, this account shows how a complete body of artistic knowledge based on rational insight into the nature of things could be compatible with failure.

VI

Of course, the Rationalists' account of medicine's stochastic character did not satisfy the Empiricists. They held the Rationalist to the promise of certainty implicit, they maintained, in his claim to real knowledge of the nature of things. And they argued that the Rationalists' own failure refuted the Rationalist account. Though he was out of sympathy with it (*Subfig. emp.* c. 9 Deichgräber), Galen does preserve a few traces of the Empiricists' anti-Rationalist polemic. He reports that the Empiricists think they have refuted the Rationalists when the Rationalists cannot respond to their challenge to give an example of a for the most part theorem which they have made certain (*adiaptotos*) and invariable (*dienekes*) by the addition of reason (*logos*) (In *Hipp. prog. comm.* CMG 9 2, 377, 24 ff. = 107, 21 ff. Deichgräber). The Empiricist spokesperson in Galen's youthful work *On Medical Experience* makes a similar charge (153 Walzer):

> As for you, if you also say that you are baffled in these matters and fall short of attaining the truth in regard to them, you prove the case against yourself. If you should say, however, you are not baffled, then pray tell us why you fail to obtain your object, since it is incumbent on you, in virtue of your self-advertised claim to possess knowledge of the paltry things even of this degree of minuteness, that you should always be correct and successful and reach your goal, as far as is humanly possible.

The Empiricists drew two conclusions from the Rationalists' failures. First, that Rational medical theory had no practical benefit to offer medicine (cf. *De sect. ingred.* SM III 10, 10-13). On this point they came into direct conflict with the Rationalists, who gave as their reason for theorizing the improved, perhaps vastly improved, ability to make correct prognoses and to treat patients successfully it was supposed to make possible. But, the argument went, the Rationalist was no better off than the Empiricist in this respect. They both relied in effect on what was for the most part true. So Rational theory was at best superfluous, since it added nothing to medicine which could not be achieved by other means. And second, the Empiricists maintained that the Rationalist position was fundamentally unsound as an account of the nature of medical knowledge because it offers no explanation for the stochastic character of medicine. Unlike the Empiricist, the Rationalist laid claim to certain and infallible knowledge. The rational inferences he employed were supposed to bestow complete certainty on their conclusions. According to the Empiricists, the result was that there was no place for uncertainty or failure in the Rationalists' account of medical knowledge. If the Rationalists' account were correct, medicine would not be a stochastic art. Therefore, the Empiricists concluded, the Rationalists are compelled by their own standards to regard failure as conclusive evidence that Rational medical theory did not meet the standards of completeness imposed on artistic knowledge.

On the other hand, the Empiricists held that their own account did not suffer from the same defects. On their view, the art of medicine is stochastic because the theorems that make up medical knowledge are themselves stochastic ([Galen] *De opt. sect.* I, 114 K). By speaking in this way they meant to call attention to the fact that Empirical theorems included an explicit specification of the relative frequency of the connection they report. Four levels of frequency were distinguished: always, for the most part, (roughly) half the time and rarely (Galen *Subfig. emp.* 45, 25-30, 58, 15 ff. Deichgräber; cf. *On Medical Experience* 95, 112 Walzer; [Galen] *Def.*

med. XIX, 354, 12 ff. K = 58 Deichgräber). Thus the Empiricists' account permitted the form of medical knowledge to reflect the fact that signs have different frequencies of correlation with the things they signify and that different treatments have different rates of success in different circumstances.

The reasons why the Empiricists took this step are not difficult to see. The investigator who doubts whether he has, or is ever likely to achieve, rationally grounded insight into the underlying nature of things must content himself with what can be gathered from experience; and he will have to take account of the fact that, unlike rational inference, which offers only one level of total evidential support, experience gives different levels of support to the conclusions it warrants. Because coincidences of observable events—which are, according to the Empirical view, the object of medical knowledge—are characterized by different frequencies, a committed Empiricist will have to make medical knowledge reflect this fact. It is at this point that we arrive at a recognizable, frequency-based conception of probability. The Empiricists' refusal to traffic in hidden natures and rational necessities made it possible for them to take a step which adherents to the Aristotelian account were unable to take. As we saw, Alexander of Aphrodisias explicitly repudiated lower levels of frequency because, according to him, truths about what occurs more or less half the time or less frequently were useless and of no interest to the artist since they lacked the privileged relation to a thing's nature that only truths about what happens always or for the most part possess (*In an. pr.* 36, 19 ff.). But because they made observed patterns of co-occurrence instead of the real underlying nature of things the object of knowledge, the Empiricists were in no position to discriminate against lower levels of frequency. Thus the innovation that opened the way for the development of a probabilistic conception of artistic knowledge was their recognition that a range of relative frequencies taken by themselves, without any reference to the underlying causes, was the proper concern of artistic knowledge.

And we can see that the Empiricists assigned a crucial role to observed relative frequencies in their account of inference generally. The probability of the conclusion of a simple sign-inference was the relative frequency of the co-occurrence of the sign and the item signified (as long as the background evidence consisted of a sufficiently large number of observations). But the Empiricists' account of the transition to the similar—their method of reasoning on the basis of experience in novel cases—also appealed to observed relative frequencies. The degree of probability or warranted expectation appropriate to the conclusion of such a transition should be proportional to the weight of the evidence that counts in its favor. (It is their willingness to talk about hope or expectation in this context which is the basis for talk of the probability of the conclusion.) And the degrees of similarity in the transition to the similar were weighted by reference to the relative frequency of successful inferences based on the type of similarity in question (*Subfig. emp.* 70-75 Deichgräber). The Empiricists' advocacy of frequency based probability grounded in observation was, however, coupled with a repudiation of the plausible (*pithanos*) or the likely (*eikos*) as they might have been used in Rationalist theory construction (Galen *De sect. ingred.* SM III, 10, 9; cf. *Subfig. emp.* 64, 31 Deichgräber). They did not see any basis for assessing the likelihood of claims which were true or false universally and invariably but nonevidently, i.e., claims about the permanent underlying nature of things which were not characterized by observable frequencies. Thus they did not concede degrees of rational likelihood to their opponents' theories. Atomic theory, e.g., though plausible to some, struck the Empiricists as the purest speculation. It was not the kind of thing to which they would be able to assign a probability, for the only source they recognized for such judgements was experience, which told them nothing about the nonevident.

It should be clear now why the Empiricists thought they did justice to the stochastic character of medicine in a way that their Rationalist opponents could not. There is no room in the Empiricists' account of medical knowledge for a gap to open up between the promise implicit in the claim to real knowledge of the underlying nature of things and failures in practice. When a certain therapy has been observed to be successful only for the most part, Empiricists will expect it to be successful in most but not all of the cases in which they make use of it. The same holds for the other degrees of observed correlation. As a result, occasional failures are an expected and predictable part of medical practice, instead of the nuisance they are for Rational theorists who must, almost as an afterthought, resort to ad hoc additions to their theory to explain them away. As long as they do not occur more often than the theorems predict, failures are not a reproach to the art of medicine or the medical practitioner. But, if an Empiricist has learned his art well, this is most unlikely to happen. Thus the Empiricists' innovative reliance on empirical probabilities made possible a very effective explanation of how failure is compatible with artistic knowledge. Their account shows how it is possible to maintain that the success an art may achieve is limited by nature without claiming to grasp the nature of the matters with which the art is concerned.

Of course, one consequence of the Empiricists' view is that it becomes harder to fix precisely the point at which an art becomes complete. Though they may correctly assert *that* nature places limits on the success of the stochastic arts, they cannot, in the Aristotelian manner, *explain* failure by referring to an account of the nature of the matters with which the arts are concerned that can be seen to be complete on independent grounds. Nor can they append to a theory of the Rationalist type, which can also be seen to be complete on independent grounds, references to the inescapable difficulties of applying the theory, developed as an explanation of why a complete body of theoretical knowledge cannot be applied in practice with complete success. Instead, they substituted the notion of a sufficient amount of experience of a wide enough variety of occurrences, experience which was composed not only of a physician's own observations (autopsy) but also of the collective experience of the medical profession preserved in the writings of trustworthy physicians (history).

VII

The debate between the Rationalists and the Empiricists did not end here, of course. In particular, the Empiricists' claim that experience could give rise to a complete body of medical knowledge was challenged by the Rationalists, who argued that experience could not provide the basis for a system of medical knowledge flexible enough to deal with the varied demands of medical practice. And the Empirical position was not obviously immune to this kind of challenge. For one thing, the Empiricists sometimes treated signs as the equivalents of symptoms (Galen *De causis continentibus* = 140, 22 Deichgräber). Symptoms, however, are individual features of a case which combine to form syndromes (Galen *Subfig. emp.* 56, 4 ff. Deichgräber). Since they refused to postulate an underlying pathology which gave rise to syndromes of symptoms, the Empiricists treated syndromes as unities only insofar as they consist of symptoms which can be frequently observed to occur together in a certain pattern. Different types of these syndromes were distinguished by reference to their evidential functions. Thus prognostic and therapeutic syndromes are so called because they point to future developments and appropriate therapies respectively (*Subfig. emp.* 58, 4 ff.; 126, 5 ff.; 143, 19 ff. Deichgräber). Diagnostic or pathognomic syndromes pose a special problem. For a Rationalist, a syndrome of manifest symptoms can be the basis of an inference to the underlying pathological condition, e.g., an imbalance of the humors. For the Empiricists, however, there is nothing more to a disease (so far as they know) than the syndrome (*Subfig. emp.* 57, 2; *De caus. cont.* = 140, 15 Deichgräber). There is no room for an inference to a new conclusion; Empirical diagnosis seems to be just a matter of applying a name (cf. Deichgräber [1930] 1965, 310). This view of the matter draws some support from reports that the Empiricists gave names to syndromes for the sake of concise instruction (*Subfig. emp.* 57, 10; *De meth. med.* X, 460, 20 K = 142, 5 Deichgräber). But it suggests that the Empiricists were locked into a rigid system of verbal stipulations; each disease was to be equated with an ordered set of symptoms, the presence of which was a necessary and sufficient condition for the presence of the disease.

And as we have seen, this furnished the Rationalists with the grounds for one of their most serious charges against the Empiricists. According to them, the Empiricist was burdened with an immense and unwieldy system of distinctions with the result that the slightest change in the symptoms, their order or their intensity required the postulation of a new syndrome ([Galen] *De opt. sect.* I, 135 K, Galen *On Medical Experience* 90 Walzer). The Rationalists argued that the welter of syndromes that resulted did not provide a perspicuous basis for therapeutic inference. The Rationalist could make use of artistic conjecture. The Empiricist, by contrast, will be dependent on so-called exact syndromes, the recommended treatment of which is always the same. But more often than not the physician is not so lucky, and the observable symptoms differ in some small way from any syndrome previously experienced and may require different treatments (Cf. Galen *De loc. aff.* VIII, 14, 7 ff. K = 85 Deichgräber; *De Sanit. tuen.* CMG V 4.2 161, 12 ff. = 161, 26 ff. Deichgräber). If this argument is right, the Empiricists' claim that experience could be the basis of a body of knowledge complete and systematic enough to constitute an art may not hold up. Empirical knowledge could turn out to be in principle incomplete and unsystematic. The fact that Empirical physicians are able to treat patients suffering from novel maladies effectively must be due to covert reliance on rational theory, for without hidden theoretical commitments the Empiricists could not group symptoms together and identify medically relevant similarities between cases.

Galen and the author(s) of the *De optima secta* (cf. Mueller 1898) are in sympathy with these charges. Unfortunately they do not tell us very much about how the Empiricists might have responded. But the Empiricists need not have conceded defeat. For one thing there is no reason to suppose that experience could only support conclusions stated in terms of the finest-grained distinctions possible. It might equally well support conclusions about kinds of items and episodes related by similarities of different degrees of generality. The Empiricists may have found that different patterns of symptoms which could be distinguished were better left undistinguished for purposes of diagnosis. The Empiricist will explain that long experience has taught his school that it is useful to group symptoms together in the way they do though more distinctions are possible. And this does not have to be done on the basis of a prior consideration of the most precise groupings possible. Experience may have shown that certain kinds of imprecise similarities provide the best basis for therapeutic inference. The Empiricists' willingness to use terms borrowed from Rationalist medicine, e.g., plethora, pleurisy and the like, suggests that they did not in practice feel any more committed to excessive precision than the Rationalists. What the Empiricists denied was that this practice implied a commitment to the pathology in terms of which the Rationalists explained their distinctions.

With these thoughts in mind, it is possible to form a somewhat more sympathetic picture of Empirical diagnosis. One clue about the Empirical conception comes from Galen's comparison of their practice to a view of the Herophileans. It seems that these doctors introduced a trichronal division of signs. They take some signs from the past, some from the present and eventually some signs from the subsequent development of the case including the patient's response to treatment. On the basis of these they draw a conclusion about what the condition is or was (Galen *De plen.* VII, 534, 13 ff. K = 155, 35-156, 19 Deichgräber). (Galen has some difficulty treating the last mentioned signs as diagnostic because they are too late to be of any use in therapy.) One possibility, then, is that the whole history of a patient's symptoms was used as the basis of an overall assessment, a backward-looking diagnosis. This would not have been of much use in the treatment of the patient when he was in need of treatment. But it could contribute to a physician's stock of knowledge and be of benefit when similar cases arose in the future. A second possibility, compatible with the first, is that the physician is engaged in diagnosis from the moment he first examines the patient onwards. At each stage he tries to take stock of the disease he is treating in order to be able to predict its future course and decide on an appropriate treatment. He will also be interested in getting a fix on what kind of disease it is he is dealing with. Some diseases will be such that a relatively small set of symptoms seen early on are enough to decide what disease it is. Sometimes it will be necessary to wait for the outcome. And in general further investigation will have to be guided by preliminary, though revisable, diagnoses. And there will be many cases that fall somewhere in between. If the Empiricists shared this conception, drawing a diagnostic conclusion did not involve waiting until all the facts were in. At different stages of the disease the Empiricist would feel different and increasing degrees of confidence in his diagnostic hypotheses about what kind of syndrome he was dealing with.

Some support for this view can be drawn from the Empiricists' willingness to talk of a difference in power among signs (*In Hipp. de medici officina comm.* XVIII B, 645, 8 ff. K = 145, 8 ff. Deichgräber). Some signs are common to many syndromes, some are common to a few and some, perhaps, belong only to one. In general, the Empiricists suggest, the fewer syndromes a symptom is part of the more powerful a sign it is. Of course, this does not say enough. It would be a mistake to suppose that every symptom bestows probability $1/n$ on each of the n hypotheses that it is one of the n possible syndromes that is involved. I suspect that the Empiricists spelled out the difference in the power of signs in much the same way that they handled the transition to the similar. The evidential value of a sign was proportional to the observed relative frequency of its occurrence in the syndrome in question and the same was true of combinations of symptoms.

One potential problem with this account concerns the level at which the observation is supposed to take place. Will the physician have to have observed (or learned through history) how often each possible combination of symptoms has turned out to be a part of the syndrome in question in order to form a diagnostic judgement in each case? In the comparable case of the transition to the similar, if one has observed that items similar in two respects of a certain kind are more often similar in respect of curative power than objects only similar in one such respect, and that objects similar in three respects of the appropriate kind are still more frequently similar in the crucial respect of curative power, it can be asked, has one observed that the more respects of the appropriate kind in which objects are similar, the more often they are similar in curative power? If so, Empirical sign- inferences may also have been less rigidly tied to the observation of precisely the symptoms in question in a particular case than it at first appears. An Empiricist will not need to have experienced the frequency with which just this set of symptoms has turned out to be the initial stage of a certain syndrome in order to form a judgement about how likely it is that it is this syndrome he is dealing with. In fact, it seems that the Empiricist will rely on something like the Rationalist's artistic conjecture. Theoretically, so to speak, their diagnostic inference will be backed up by observation of the tendency of certain types of symptoms to combine in certain ways with more frequency than in others. This is so even when the particular combination that is the basis of his diagnosis in the case in question is not one he has experience of or has learned about in the literature. For in practice he will be guided by expectations which cannot be justified with perfect precision. Thus his view will involve an element of idealization just as the Rationalist's view that his conjectures are guided by rational theory does.

The debate did not end here of course. The Empiricist continued to press the Rationalist to explain how the rationally warranted insights to which he laid claim permitted failure, while the Rationalist argued that the empirical resources on which the Empiricist claimed to rely exclusively are not adequate to explain his successes. But it should now be clear in how many ways the concern to account for failure affected ancient conceptions of artistic knowledge.

University of Pittsburgh

Notes

*I am grateful to Michael Frede, Katharina Kaiser, Geoffrey Lloyd, Heda S̄ egvić, Peter Singer, Steven Strange, the participants in the Workshop on Scientific Failure, University of Pittsburgh, April 1988 and the members of the Society for Ancient Greek Philosophy in attendance at the 28 April 1989 meeting for their very helpful comments on earlier versions of this paper.

1. I am grateful to Terence Irwin for calling this point to my attention.

2. Of course, natural developments, e.g., in the life history of a living thing, which do not take place for the most part, at least under a certain description, need not be exceptions to a for the most part pattern of developments in accord with the nature of thing. Some matters are underdetermined by the nature of the thing at issue. Thus the

particular color of a human being's eyes is not specified by that person's human nature, though the range of colors which will permit the eye to fulfill its function may be, so that it will be because of the nature of a human being that the color of human beings' eyes for the most part fall somewhere within that range (cf. GA 778a32-b1).

3. I have departed from Kuehn's text according to which both stochastic medicine and the medicine of Apollo and Asclepius are called "true medicine." This undermines the contrast which is being drawn. For this reason, I believe the first occurrence of "true" ought to be bracketed thus: ἡ [ὀ'ντωδ] ἰατρική.

References

Collections

Deichgräber, K. (1965), *Die griechische Empirikerschule: Sammlung der Fragmente und Darstellung der Lehre*, 2d ed. Berlin: Weidmann.
Diels, H. (1929), *Doxographi Graeci*, Berlin: De Gruyter.
Radermacher, L. (1951), "Artium Scriptores (Reste der voraristotelischen Rhetorik)" *Sitzungsberichte der philosophisch-historischen Klasse der Oesterreichischen Akademie der Wissenschaften*, 227 band 3.

Classical Works

Alexander Aphrodisiensis (1883), *In Aristotelis analyticorum priorum librum* I, ed. M. Wallies, Berlin: Reimer (Commentaria in Aristotelem Graeca II, 1).
_____. (1892), *Quaestiones*, ed. I. Bruns, Berlin: Reimer (CAG Suppl. II, 2).
_____. (1891), *In Aristotelis Topicorum libros*, ed. M. Wallies, Berlin: Reimer (CAG II, 2).
Galenus (1821-33; repr. 1965), *Opera quae exstant*, ed. K. G. Kuehn, Leipzig; Hildesheim. (K)
_____. (1884-93), *Scripta minora*, eds. G. Helmreich, J. Marquardt, I. von Mueller, Leipzig: Teubner. (SM)
_____. (1969), *De causis continentibus et al.*, ed. M. Lyons et al., Berlin. (Corpus Medicorum Graecorum Suppl. Orientale II)
_____. (1923), *De sanitate tuenda*, ed. K. Koch, Berlin. (CMG IV 2)
_____. (1944), *On Empirical Medicine*, ed. and trans. R. Walzer, Oxford: Oxford University Press.
Philodemus (1892-96), *Rhetorica volumina*, ed. S. Sudhaus, Leipzig: Teubner.
Sextus Empiricus (1914-58), *Opera*, eds. H. Mutschmann, J. Mau, Leipzig: Teubner.

Modern Works

Deichgräber, K. (1956), "Galen als Erforscher des menschlichen Pulses: ein Beitrag zur Darstellung des Wissenschaftlers," *Sitzungsber. d. Dt. Akad. d.*

Wiss. 35, 3-47, reprinted in K. Deichgräber, *Ausgewahlte kleine Schriften* G. Gartner *et al.* (eds.), Berlin: Weidmann, pp. 288-326.

Edelstein, L. (1967), "The Hippocratic Physician," in L. Edelstein *Ancient Medicine*, O. Temkin (ed.), Baltimore: John Hopkins, pp. 87-110.

Heinimann, F. (1961), "Eine vorplatonische Theorie der Technē," *Museum Helveticum 18*: 105-130.

Mueller, I. von (1898), "Über die dem Galen zugeshcriebene Abhandlung Περί τῆ δ ἀ ρί στη δ ἀ ιρέ σεωδ," *Sitzungsber, der philosophischen-philologischen Klasse der k. b. Akad. d. Wiss.*, pp. 53-162.

6. SCHOLASTICISM AND THE PHILOSOPHY OF MIND: THE FAILURE OF ARISTOTELIAN PSYCHOLOGY*

Peter King

1. INTRODUCTION

There are many kinds of scientific failure such as experimental results which are incompatible with or unexplained by scientific theory; the specific invalidation of a given hypothesis; and the abandonment of a promising theory. Yet these are not the kinds of failure I will address. Rather, I am interested in the most general kind of scientific failure, namely, the collapse of a research program. A research program need not collapse when a scientific theory is discarded, for the failure of a particular theory may not invalidate the general approach which the particular theory embodies: the nexus of common assumptions, the method of exploration and validation, the promising lines of development and research, even the very terms in which the debate is couched, may well all survive the demise of a particular theory. These features characterize a given research program—or "scientific paradigm"—and their persistence typifies a (reasonably) unified scientific tradition. Conversely, substantial changes in these features, or their wholesale abandonment, mark the failure of a research program.

The failed research program with which I will be concerned is the mediaeval articulation and development of Aristotelian psychology. Its failure is instructive and complex. Briefly, I argue for the following theses: (i) the mediaeval paradigm for psychology was such that it generated an insoluble problem, namely, what I call the "problem of transduction"; (ii) that the failure to resolve this problem was instrumental in the eventual abandonment of the mediaeval paradigm itself; (iii) that its successor, Cartesian psychology, is directly indebted to the collapse of the mediaeval paradigm. And apart from the historical argument of (i)-(iii), the failure of Aristotelian psychology is interesting, for the problem of transduction— and the features of the mediaeval research program which led to its formulation (and perhaps its insolubility)—again occupies center stage, in the burgeoning field of cognitive science.

What psychological mechanisms, functionally defined, have to be postulated to account for the facts of mental life? The contemporary ring to this question is due to its prominence in recent work in cognitive science and the philosophy of mind—largely spurred by issues in the philosophy of psychology and artificial intelligence. Yet the very same question could have been asked with equal propriety during the heyday of the "Aristotelian revolution" in mediaeval philosophy, the century of High Scholasticism (1250-1350), which concentrated on issues pertaining to "mental architecture." According to the standard Aristotelian analysis, the soul possesses cognitive faculties, that is, sensitive and intellective capacities. Less grandly put, they could feel and think with their souls, just as we do with our minds—and that is the question to be discussed here: whether mediaeval philosophers found it necessary to postulate a psychological mechanism mediating the cognitive faculties of sense and intellect, and, if so, how such a mechanism functions. The inability of the Scholastic tradition to reach consensus on a response to this question eventually led to the wholesale collapse of the Aristotelian approach to the mind.[1]

I proceed as follows: Section 2 sets forth the elements of the problem of transduction; Section 3 canvasses the common mediaeval understanding of Aristotelian psychology; Section 4 is devoted to mediaeval transductive accounts of understanding, namely, abstractive theories and illuminative theories; Section 5 discusses the rejection of transductive mechanisms and the problems which arise from such a rejection; Section 6 turns to Cartesian psychology and its central theses in the context of the mediaeval agenda.

2. THE PROBLEM OF TRANSDUCTION

Are there special conditions which a psychological mechanism mediating the cognitive faculties of sense and intellect must satisfy? Zenon Pylyshyn (1984, chap. 6, esp. 153-178) has recently argued that there are; such a mechanism must be what he calls a "transducer," which can be described as follows:[2]

- A transducer is, roughly, a stimulus-bound mechanism which is data-driven by its environment, operating independently of the cognitive system.

- The behavior of a transducer is to be described as a function from physical events onto symbols.

- The function carried out by the transducer is primitive and is itself nonsymbolic. At least, the function is not described as carried out by means of symbol processing; it is part of the functional architecture of the mind.

A transducer is a stimulus-bound mechanism in that the input for its activity derives from environmental rather than cognitive sources; as part of the functional architecture, it is at least relatively independent of cognitive processes—susceptible to gross influences such as changing the direction of one's gaze, but not altered by changes in cognitive states such as beliefs or desires. It is data-driven by the environment in that the input, in this case states of the sense-organ(s), is modified by the environment. A transducer, then, is a psychological mechanism which is "cognitively impenetrable." To mediate between sense and intellect, a transducer must map physical input, such as the deliverances of the senses specified physiologically, onto output which is "intellectual" in nature. A minimal condition for being "intellectual" is that the output be describable symbolically: roughly, that it be language-like at some level, producing as output tokens which may then be susceptible to rule-governed manipulations, as words are grammatically combined into sentences. Yet the function which the transducer performs must itself not presuppose any "symbolic" operations; it is useless to try to explain the transformation of the elements of sense into elements of the intellect by presuming some form of intellectual operation involved in the transformation. This is precisely the "homunculus" mistake in the philosophy of mind. Finally, the function accomplished by a transducer is primitive with regard to the rest of the cognitive system (a "one-step" process): it performs a single operation with no internal cognitive steps.

The question at hand, then, is whether a psychological mechanism meeting these requirements can be found to mediate between the faculties of sense and intellect, which I call *the problem of transduction*. Three principles generally accepted in the Scholastic period made the problem of transduction pressing and acute: (i) the difference between sensing and understanding is a distinction in kind, based on the difference between the faculties of sense and intellect (this is not to say that the sensitive soul and the intellective soul are in some sense really distinct entities, although this is the customary mediaeval position; a single entity may have qualitatively different features; see f.n. 7); (ii) understanding may be characterized linguistically, so that concepts are thought of as (literally) mental words;[3] (iii) the intellect is initially a *tabula non scripta*, so that its mental "vocabulary" must be acquired. The conclusion typically drawn from (i)-(iii) is that the "words" making up the intellect's "vocabulary" are somehow derived from, or intimately related to, sense—that is, that there must be a transductive mechanism.

Of those mediaeval philosophers who accepted the need to posit a transducer, two general accounts of the transducer's function predominated: One group held that the transducer is abstractive, the other that it is illuminative. (The debates in the Scholastic period often concerned questions about the transducer's nature as well, specifically whether it was a faculty possessed by each individual soul, a suprapersonal single faculty, or in some sense divine—that is, the controversy with so-called "Latin Averroism." I concentrate on the transducer's function, ignoring such other issues.) Those who took the transducer to be abstractive, such as Thomas Aquinas and Duns Scotus, argued that the elements of the sensitive soul are literally taken up and transformed; those who took the transducer to be illuminative, such as Bonaventure, Matthew of Acquasparta, and Henry of Ghent, argued that the elements of the sensitive soul are not themselves operated on but rather viewed in a new light. Other philosophers, such as Peter John Olivi, Durand of St.-Pourçain, and William of Ockham, rejected the need to posit a transducer (or a specific and identifiable mechanism for transduction), arguing instead that no such mechanism is necessary.

These philosophers were all in some sense "Aristotelians"; at least, their scientific research was carried out against the background of Aristotelian philosophy, even when Aristotle's own analysis was rejected. Therefore, a preliminary discussion of the Aristotelian "science of the mind" will set the stage for the debates over the problem of transduction during High Scholasticism. First, however, some caveats are in order. Concentrating on a single problem will inevitably distort the actual historical development of positions: by concentrating on mental architecture, I will put aside epistemological worries about knowledge and its justification, which often motivated the debates; by concentrating on transduction, I will put aside issues having to do with, for example, perceptual illusion and the physiology of perception. Yet the problem of transduction is interesting in its own right, and organizing the vast quantities of mediaeval literature around this problem will allow certain thematic developments and positions to stand out more clearly than they might otherwise. Note finally that the version of Aristotle I present will necessarily be simplified, and indeed I hold no brief for it being Aristotle: It is rather the common mediaeval reading of Aristotle.[4]

3. ARISTOTELIAN PHILOSOPHY OF MIND

For the Aristotelian, the distinction between the living and the nonliving is a matter of the presence of "soul" (*anima* or ψυχή): an entity postulated to explain obvious differences between the living and the nonliving, such as understanding, sensing, local movement, nutrition, growth and decay.[5] Living beings, composites of body and soul, are paradigmatically *things*— beings structured and governed by internal principles, unlike, say, mud, fingernails, or the conjunction of my left earlobe and the dark side of the moon. The unity of soul and body is the tightest possible in Aristotelianism: soul and body are related as form and matter, where the soul is the form of the body in a literal sense, as that which informs the (merely) physical organic construction which is the body and makes it be the kind of biological unity it is. Just as a given shape organizes a lump of bronze into a statue, so too soul organizes a lump of bodily organs into a living being. Form and matter are generally related as act and potency; the given shape makes the lump of bronze, only potentially a statue, into an actual statue; the shape actualizes the potentialities of the bronze in a determinate way—the shape makes a potential statue (the lump of bronze) into an actual statue, and so may be called the "actuality" of the statue. Soul, likewise, is the actuality of body.[6] The potentialities of body which soul actualizes—nutrition, growth, movement, sensing, understanding—are present only in a hierarchy: bodies which sense assimilate nutrition, but not conversely. There seem to be three "kinds" of soul, that is, clusters of principles which are actualized: (i) nutrition and growth, as in plants; (ii) sensing and movement, as in brute animals; (iii) understanding, as in humans. These are known respectively as the vegetative soul, the sensitive soul, and the intellective soul.[7]

Aristotelian philosophy of mind is constructed around the following central principle: Understanding is to be thought of after an analogy with sensing.[8] Thus the analysis of the sensitive soul, itself based on an analogy, will provide the key to the intellective soul. Aristotle describes the process of sensing as follows:

> Sense is that which has the power to receive into itself the forms of sensible objects without the matter, just as a piece of wax receives the impression of the signet-ring without the iron or gold [of the ring itself]; what produces the impression is the iron or gold [signet-ring], but not as iron or gold. In a similar way, sense is affected by what is colored or flavored or sounding, but not by what the substance of each of these is; rather, only as having a certain quality, and in virtue of its definition. (De anima II.xii 424a 17-24)

An external object causally acts on the sense-organ, such as the eye, putting it in a new physical state.[9] Each particular sense-organ corresponds to a particular sense-faculty; the eye is the sense-organ of the faculty of vision, the ear the sense-organ of the faculty of hearing, and so on. In general, the sense-faculty is the form of the sense-organ—a particular instance of the form-matter relation between soul and body.[10] The signet-ring leaves an impression on a piece of sealing wax; analogously, the external object acts on the sense-organ to leave in the sense-faculty an "impression." Three points of comparison stand out:

- The sealing wax itself, while possessed of a determinate nature ("waxhood"), can take on many different physical configurations; it can be stretched, shaped, and so on while still remaining wax. The limits of these possible configurations are determined by its nature.

- When acted upon by the signet ring, the sealing wax takes on a determinate configuration; it becomes something new, the "composite" entity which is the seal. Different seals correspond to different configurations.

- The sealing wax takes on formal features of the signet-ring, the shape of the seal, but not the material features; the iron or gold of the signet-ring is "left behind."

Wax can be in different physical states due to its malleability and ductility, which are part of the nature of wax. The sense-organ, analogously, has a determinate (organic) nature, and its ability to be in distinct physical states is due to the organ being animated—that is, being the material organ of a given sense-faculty, which must be part of a living being; the rods and cones of a corpse's eye do not register the effects of light. While the actual material structure of the eye is part of its organhood, the reactivity of the eye, its receptivity to causal affection, is due to the animated nature of the sense-organ. The sense-faculty, in conjunction with the material composition of the organ, determines the possible physical states the sense-organ may occupy. The sense-faculty is potentially any of the admissible physical states of the organ, as the sealing wax is potentially any seal. When the sense-organ is put into a new physical state by the causal action of the external object, a "composite" is formed, as when the wax is impressed by the signet-ring, the composite entity which is the seal is the result. The signet-ring actualizes some of the potencies inherent in the nature of wax so that it becomes determinately wax which is a seal, or wax which is the matter of the seal. In the case of the sense-organ, the new "composite" is the *sensing of the object*. The external object actualizes some of the potencies in the sense-faculty so that it becomes determinately a sensing of the object, and the sense-organ in the given physical state is the matter of the sensing. The state of the sense-organ and the sensing are one and the same in just the way that the matter of something and that of which it is the matter are one, namely, by being the determinate actualization of a potency.[11]

Now an iron or gold signet-ring impresses its formal structure (a geometrical pattern) on the sealing wax, but without its matter: The sealing wax remains wax rather than iron or gold, and it is "one" with the impressed pattern, the formal structure of the signet-ring, as a determinate actualization of a potency. The seal *is* the embodiment of the formal structure of the signet-ring in wax. Analogously, the sensible object impresses its formal structure on the material sense-organ, but without the matter. The sense-faculty formally becomes "one" with the object, that is, takes on the form of the object—not merely similar features, but identically the same form. The sensing of an object embodies the form of the object in the sense-organ: It formally *is* the object. The form in the external object inheres in matter, and so makes the object to be the very object or the kind of object it is (say, a sheep); the form as inherent in the sense-faculty does not make the sense-faculty into a sheep, but into a sensing-of-a-sheep. It is one and the same form in both cases, differing only in the mode of inherence.[12]

Sensing is of an object, not of a form (whether the form in the soul or the form in the object)—a fact which immediately leads to a complication. The wax is not molded into the exact shape of the entire signet-ring, so that in addition to the iron ring there is a wax one; rather, the wax takes on one aspect of the formal structure of the signet ring, namely, the geometric pattern on the face of the ring. The wax is formally identical to the facing shape of the ring, not the ring itself. Analogously, in sensing, the sense-organ takes on an aspect of the formal structure of the object. The faculty of vision is formally identical to the visible elements of the formal structure of the object, the faculty of touch is formally identical to the tactile elements of the formal structure of the object, and so on for each of the five external senses. Now the distinct sense-modalities are usually referred to one and the same object: It is the sheep which, for example, looks, feels, and smells, a certain way. To account for this unity, an "internal sense" parallel to the five external senses is postulated, called the "common sense," which unifies the distinct sense-modalities.[13] The common sense functions in exactly the way the external senses do: It takes on the various modal forms reported by each of the senses, and is put into a determinate physical configuration. The receptivity of the commonsense faculty to being put in such configurations is a matter of its being animated; the determinate actualization of the common sense's potencies is called the "sensible species."[14] Therefore, the sensible species, which is the product of the common sense, includes the totality of the object "for sense": It unites the three-dimensional colored expanse, the single pungent odor, and so on, into an object—the sheep. (Note that the sensible species does not distinguish between a sheep and a wolf which is in sheep's clothing, wearing *eau de mouton* perfume, and the like. Nor should it; we can be fooled by imitations. Nothing in the Aristotelian theory insists on the "veridicality" of sensing. The mediaeval development of theories of "intuitive cognition" will bring this point to the fore.)

To summarize; the Aristotelian analysis of sensing turns first on an exact understanding of the form-matter relation of the sense-faculty to its associated sense-organ, and then on treating this relation as a variety of the act-potency relation. The object and the sensing are formally identical. The sense-faculty is merely passive[15] to begin with, and is only potentially its objects (formally speaking). In general, something is reduced from potency to act only by an agent cause, that is, whenever there is some actualizing process going on there is an agent which causes the occurrence of that process.[16] The sensed object is the agent cause of the determinate actualization of the potencies of the sense-faculty. External objects are actually sensible; in standard circumstances, they causally bring it about that they are actually sensed. The distinction of external and internal senses seems required by the evident facts of experience, but each faculty is given the same kind of analysis.

Therefore, Aristotelian philosophy of mind endorses a potency-act-cause analysis of sensing. Since understanding is analogous to sensing, it too will be given a potency-act-cause analysis. Understanding, like sensing, is a process of taking on the form of the object:

> The intellect, although impassible, must be receptive of the form of the object, that is, it must be potentially the same as its object without being the object: as the sensitive is to what is sensible, so too the intellect to what is intelligible. (*De anima* III.iv 429a 16-20)

> Actual understanding is identical with its object. (*De anima* III.v 430a 22-23)

Just as the sense-faculty takes on formal features of the external object, the intellect too takes on formal features of the same object. The faculty in the intellective soul which is passive and receptive (of the form of the object) is called the "possible intellect" or the "material intellect." The reception of the form of the object determinately actualizes the intellect, previously only potentially the same as the object, such that the intellect is actually identical with the object (formally speaking). When the intellect takes on a form and so is determinately reduced to act, it becomes a *thinking of the object.*[17]

Thus far everyone is in agreement. Yet since nothing is reduced from potency to act without an agent cause, and the intellect is only potentially the same as its object, there must be an agent cause of understanding. The difficulties and disagreements arise with regard to identifying (i) the "form" taken on by the intellect; and (ii) the agent cause of understanding. Aristotle's remarks which purport to address (ii) are famous for their obscurity:

> The intellect as we have described it is what it is because it becomes all things. There is another which is what it is because it makes all things: this is a kind of positive state like light, for in a way light makes potential colors into actual colors; the intellect in this way is separable (χωριστὸς), impassible, and unmixed, since it is essentially an activity—for the active factor is always superior to the passive factor, and the originating cause [is always superior] to the matter. (*De anima* III.v. 430a 14-19)

The possible intellect "becomes all things": it is the "form of forms," taking on any form of any object. But the sensible species produced by the common sense are only potentially intelligible, just as colors are only potentially seeable until light shines on them. According to one interpretation of this passage, the phrase "there is another" picks out an intellective faculty distinct from the possible intellect, called the "agent intellect," which transduces the potentially intelligible into the actually intelligible, as light transforms the potentially seeable into the actually seeable. The agent intellect, then, acts as the "light of the mind." According to another interpretation, "there is another" picks out another way of describing the activity of the intellect, and the transduction of the potentially intelligible is a primitive and indivisible function of the intellective soul itself. Philosophers who adopt the first interpretation offer transductive accounts of understanding; those who do not, offer nontransductive accounts.

4. TRANSDUCTIVE ACCOUNTS OF UNDERSTANDING

4.1 Abstractive Transduction.

Thomas Aquinas and Duns Scotus, for all their differences, agree that (i) there is an agent intellect which is a faculty distinct from the possible intellect, and (ii) the function of the agent intellect is primarily abstractive. With regard to (i), Scotus offers a straightforward argument for the claim that there is an active principle of understanding in the intellective soul: It is an evident fact of experience that we can understand something not previously understood, and, as this is a nonrelational property (an "absolute form") in the possible intellect, it must be the result of some action; hence there is an active principle which brings it into existence.[18] Aquinas offers a somewhat different argument: Since the forms of material objects (given in the sensible species or the phantasm) are only potentially and not actually intelligible, there must be an active principle which makes them actually intelligible, and this reduction from potency to act requires an agent cause—the agent intellect (Thomas Aquinas, *Summa theologiae* Ia q. 79 art. 3; see also *Summa contra gentiles* II.lxxvii; *De spiritualibus creaturis* art. 9; *Compendium theologiae* c.lxxxiii; *Quaestiones de anima* art. 4; *In De anima* III lect. 10). Note that on Aquinas's view the agent intellect has two distinct and logically sequential functions: (a) preparing the sensible species so that it is actually intelligible; and (b) "impressing" this prepared sensible species, called the "intelligible species," on the possible intellect.[19]

The key premise in Aquinas's argument is that the forms of material objects are only potentially and not actually intelligible, suggested in the analogy with colors noted previously; he justifies this premise by taking the intelligible species to consist in the universal formal features of the object—which, of course, are not actually intelligible since they are not apparent to sense. (The claim that the forms of material objects are only potentially and not actually intelligible is ultimately taken from Aristotle: see *Metaphysics* II.iv 994b 18 and VII.iii 1043b 19. Equally, there is solid textual evidence in Aristotle that the intelligible species corresponds to the universal features of the object; see, e.g., *De anima* II.v 417b 23-25.) Scotus also endorses this conclusion, saying that "the agent intellect produces the universal from the non-universal...since the universal as universal does not exist" (*Ordinatio* I d. 3 pars tertia q. 1 n. 360).[20] Thus the sense has as its medium the sensible species, which is particular, and the intellect has as its medium the intelligible species, which is universal. The agent intellect is the transducer, operating prior to any occurrent thinking, which turns the sensible species into the intelligible species.

This transduction takes place through *abstraction*. The universal form is said by Aquinas and Scotus to be "abstracted" from the particular sensible species, by the removal of its individuating conditions.[21] Aquinas directly identifies the material conditions of the form as its individuating principles, while Scotus does not specify, but we can bypass these details here.[22] The elements of sense are transduced by abstraction into the building blocks of the intellect's "vocabulary," attaining linguistic character in the process. Therefore, a closer look at abstractive transduction is in order.

Despite extreme differences in their respective underlying metaphysics, both Aquinas and Scotus agree that individuating conditions are not formal differences: They do not alter the formal content of the nature of the object which they individuate, but merely render it singular, distinct from others of the same kind; formal differences only occur at the specific and generic levels. Hence the process of abstraction does not formally alter the nature, but simply removes or cancels its surrounding individuating conditions. Yet since the individuating conditions do not alter the content of the form in the individual, the form in itself must have the "abstracted" features, that is, the characteristics revealed through abstraction, though in combination with the appropriate principle of individuation the form is individualized in the object: The form in itself is "universal."[23] Aquinas and Scotus offer subtle metaphysical explanations for how the form in the individual can be individualized and yet universal in itself, and this is not the place to pursue the issue of the adequacy of their explanations; let us take it for granted, as they each did, that some satisfactory account can be offered. Now since the agent intellect operates on the sensible species and not on the object itself, the form as present in the sensible species must be universal in itself though individualized in the sensible species, in a manner analogous to the way in which the form is universal in itself though individualized in the object. The individuating conditions from which the form is released must be conditions present in the sensible species. The transductive function of the agent intellect, then, is to remove the individuating conditions from the form as present in the sensible species.

If this account, common to Aquinas and Scotus, is an accurate (though general) description of their position, the agent intellect *cannot* be a transducer, because the function it carries out is symbolic and not primitive. More exactly, the distinction between sensing and understanding cannot be maintained since the faculty of sense must have recourse to conceptual categories at a level from which they have been excluded.[24] To see why this should be so, let us consider the conclusion of the preceding paragraph, that the agent intellect removes the individuating conditions from the form as it is present in the sensible species. In order to do so, sensing itself must be classificatory, that is, the act of sensing must structure the content of what is sensed: Objects are sensed as *being of a kind*. The content of what is sensed can be represented as "this *F*," where the "this" represents whatever the individuating conditions may be, and "*F*" is a general sortal term giving the natural kind under which the thing falls. But classificatory sensing, structuring what is sensed in this way, presupposes access to general terms—to conceptual categories which brute animals are not supposed to have. The form as classified in sensing already has all of the "abstract" features required for understanding, which requires there to be conceptual abilities in the sensitive soul.

This point can be made more sharply by distinguishing the kind of classificatory sensing the theory of abstraction presupposes from both differential response and sensing what something is like. The former, differential response, does not require conceptual classification; thermostats as well as sheep respond differentially to changes in temperature; it is merely a "hardware" instantiation of the instruction "in *S* do *A*." The latter, sensing what something is like, is a matter of "being acquainted" with something, such as wolves. The sheep is "acquainted" with wolves, responding to the presence of a wolf with fear, without classifying the wolf as something which belongs to a given (natural) kind.[25] The sheep responds differentially to members of different natural kinds, but this is not to be conflated with responding differentially to them *as* members of different natural kinds; the sheep responds differentially to blue, and is even "acquainted" with the phenomenal feel of blue, without classifying blue as a color, a species different from green, and the like. The sheep need not even "sense" the wolf as an animal, much less as an individualized case of wolfhood. Yet for abstractive transduction to perform as advertised, the sensible species must contain such information, such that what is sensed is an object as a member of a natural kind. But this requires conceptualization— the concept "wolf" and the associated concepts of "natural kind" and "membership." Hence there must be conceptual abilities present in the sensitive soul, and so the agent intellect cannot be a transductive mechanism at all.

Furthermore, if the agent intellect simply removes the individualizing conditions, the essence of the kind, given in the form, must already be determinately present in the sensible species. This renders abstractive transduction even more problematic: Presumably it is part of the essence of the wolf that it is an irrational animal, yet "irrationality" is not on a par with colored expanses, discrete tastes, and the like, which are what the common sense unifies in the sensible species.[26] Nor can we "observe" the wolf's behavior and note that it does not exhibit rationality; the question at issue is how irrationality could be included in the sensible species for the purposes of abstraction, which takes place prior to any thinking—and to "note" that the wolf's behavior is irrational is an act of thinking.

Abstraction, therefore, does not provide a solution to the problem of transduction. However, an alternate account of the transducer's function was offered by other philosophers, designed to overcome these and other difficulties with abstraction: illumination theories.

4.2 Illuminative Transduction.

Aristotle said that things become intelligible through the activity of "a kind of positive state, like light"; the theories of "illumination" presented by Bonaventure, Matthew of Acquasparta, and Henry of Ghent attempt to cash out this metaphorical description. (There were also other reasons, largely theological, for attempting to develop the "light" metaphor.) The common thread uniting their theories is the claim that the elements involved in understanding are *not* present in the sensible species, however inchoate, but are contributed by the transductive mechanism itself. Transduction is accomplished when the agent intellect is guided by the Divine Ideas, which are the ideal patterns or archetypes in God's mind—they are *exemplars* (or exemplary forms) of mundane objects.[27] The exemplar explains why the mundane object is what it is, and so "illuminates" the mundane thing: The exemplar is the actually intelligible structure of the mundane object. Thus illuminative transduction takes place when the exemplar of the form in the object plays a role in the process of understanding.

Two features of this account deserve further mention. First, the relation between exemplar and concrete form is not precisely that of instantiation since the concrete form falls short of the ideal character of the exemplar. Imperfect circles—the only kind found in the mundane world—are neither perfect circles nor instances of perfect circles; they are what they are, namely, circular, in virtue of participating in the exemplar. There is no ground in the concrete form itself for circularity. Therefore, abstraction cannot serve as the logical basis for transduction. Second, this account relies on God's activity in making the Divine Ideas available for use in transduction, and so transduction is not merely a local matter of a single individual's mental architecture—but it is no less transductive for all that; mentation need not be a local phenomenon.

Illumination theories have to explain what the "activity" of the agent intellect consists in and how exemplars function in the process of understanding. On this score, Bonaventure's theory is not very enlightening; he takes the activity of the agent intellect to be the abstraction of an intelligible species from the sensible species, followed by a double impression on the possible intellect of the abstracted intelligible species (called the "created exemplar") with the Divine Idea (called the "uncreated exemplar") to produce understanding, which "co-intuits" the created and uncreated exemplars, though the latter only obscurely. This inherits all the difficulties with abstractive transduction, to say nothing of the apparent incompatibility between abstraction and exemplarism.[28] For these reasons, Matthew of Acquasparta and Henry of Ghent reject abstraction, and provide an alternative account of the activity of the agent intellect.

Matthew of Acquasparta holds that the process of abstraction is unnecessary since forms in themselves are not individualized. The sensible species produced by the common sense is a necessary, but not sufficient, condition for understanding: The agent intellect and the exemplar are partial co-causes of the intelligible species, which the agent intellect then impresses on the possible intellect.[29] In Scholastic terminology, the formal cause of the intelligible species is not the form in the object or the sensible species, as in abstractive transduction; rather, the agent intellect and the exemplar jointly constitute the formal cause of the intelligible species.[30] We understand by means of the exemplar, as we see by means of light. The color of the object and the light by which we see the color are partial co-causes of sight.

Henry of Ghent elaborated these themes to an even greater degree. The agent intellect retains the sensible species in memory as something less fixed and definite, and hence less particular; they are called "universal phantasms" for this reason—not because they present the essence, but because they do not definitely present an individual.[31] In so doing, the exemplar directly actualizes the possible intellect. Note that there is no call for an intelligible species; the exemplar takes its role, and the exemplar rather than the agent intellect acts on the possible intellect (see esp. *Quodlibeta* IX 1. 15 and, to a lesser extent, III q. 8 and IV 1. 9; some traces of this doctrine are present in *Summae quaestionum ordinariarum* art. 58 q. 2). The activity of the exemplar is due to God's agency; God is even called a kind of "second agent intellect" (*Quodlibeta* IX q. 15; this suggestion also appears in Roger Marston's *Quaestiones disputatae*). Unlike Bonaventure, in which the created and uncreated exemplars are co-intuited, and unlike Matthew of Acquasparta, in which the exemplar and the agent intellect jointly produce an intelligible species, Henry of Ghent finally came to see that the intermediate stages and doubling of exemplars was an unnecessary complication—that illuminative transduction does not require more than God's agency through the exemplar.[32] The final product of illuminative transduction is the sensible species or its generalized form in the imagination taken with respect to an exemplar.

Particular cases of illumination, therefore, are a matter of taking a sensible species (or its generalized form—hereafter I drop the reminder) *as* exemplifying, although imperfectly, a natural kind or divine pattern. The sensible species is naturally present in the sensitive soul as a matter of the Aristotelian mechanics described in Section 3. In contrast to abstractive transduction, illuminative transduction need not assume that the "informational content" of the sensible species has internal structure. To keep with the original metaphor, the sensible species is seen "in a new light," *as* presenting further information. According to the theory of illumination, there is classification taking place—insofar as taking something *as* exemplifying a divine pattern is "classificatory"—but it is the work of the intellect rather than the senses. The sensible species is like the famous "duck-rabbit" drawing: The drawing is what it is, but it may be seen as a duck or as a rabbit, and as to which the drawing is seen as is due to the intellect and not the drawing. Thus illumination avoids the trap in which abstractive transduction was caught, namely, presupposing recourse to conceptual abilities at the level of the sensitive soul.

However, illumination theories as described do not satisfy the requirements for transduction. A transducer is supposed to function without any "symbol processing," that is, to perform without presupposing any conceptual capacities. Yet precisely *which* exemplar a given sensible species is taken to exemplify has nothing to do with the sensible species itself, but rather depends on God's agency in allowing the appropriate exemplar to have causal influence in illumination. And this depends, ultimately, on God's recognition that the sensible species is of a given kind, or caused by a given kind of object, and granting causal power to the appropriate exemplar. But God's recognition is a *cognition*, and so itself involves conceptual capacities. A homunculus is no better for being divine and omnipotent, after all, and on this account God acts as a kind of "super-homunculus." Since the sensible species is not assumed to be structured in any way, nothing short of intellectual recognition on God's part can guarantee equiformity between the object sensed and the exemplar reducing the mind to act. (Henry of Ghent offers the argument in this sentence as an argument *for* illumination; see his *Summae quaestionum ordinariarum* art. 1 q. 1 *ad* 7.) Thus illumination theories covertly appeal to intellectual agency, and so fail to provide a transductive mechanism; as Henry of Ghent says, God is the "hidden understanding" (*Quodlibeta* IX q. 15) operating within us.[33]

The appeal to divine agency might be thought less objectionable if the "agency" in question were sufficiently general, that is, if God does not directly intervene in each case of understanding (called "special illumination"), but rather structures the world, or perhaps only structures the human intellect, so that the appropriate exemplar is active in the presence of the given sensible species (called "general illumination").[34] But two distinct projects have to be distinguished, namely, describing the function of a transducer and offering an account of what it is to understand. General illumination provides the latter at the expense of the former: Understanding is analyzed in terms of the subsumption of a given sensible species under an exemplar, but there is no account of the transductive mechanism because the analysis of understanding does not provide a *mechanism* at all—there is no link between the sensible species and the exemplar other than that provided by divine providence. The intellect is as passive as the senses, each faculty merely receiving the causal efflux of causes external to itself and responding in determinate ways; the "activity" of the agent intellect has evaporated. Bonaventure allowed the agent intellect real activity, but only at the cost of accepting a theory of abstraction with all its associated problems. Matthew of Acquasparta asserted that the agent intellect cooperates with the exemplar in producing the intelligible species, but does not offer any account of how the agent intellect achieves this cooperation. The agent intellect does have the function of impressing the exemplar on the possible intellect, but, in the absence of any determinate function for the agent intellect to perform on its own, there seems to be no need to retain this vestigial function; the exemplar could inform the possible intellect directly, as Henry of Ghent realized. It is but a short step to giving up the agent intellect entirely; Henry restricts the agent intellect to performing functions Aquinas and Scotus had simply ascribed to sense—retaining generalized images in memory. Thus special illumination fails to be transductive by offering a mechanism which includes an illicit appeal to agency; general illumination offers no transductive mechanism at all.

Neither abstraction nor illumination can provide any satisfactory account of transduction. Other philosophers dispensed with transductive mechanisms altogether, taking the distinction between sensing and understanding not to be mediated by any mechanism—a kind of "illumination naturalized."

5. NONTRANSDUCTIVE ACCOUNTS

Aristotle's description of the intellect does not entail that the agent intellect is really distinct from the possible intellect, and the investigation of illumination made the agent intellect seem superfluous. Accordingly, William of Ockham denied that the agent intellect is a separate intellective faculty, claiming that the agent intellect and the possible intellect are really one and only distinct in reason.[35] With the elimination of the agent intellect, there was no reason to retain the apparatus of cognitive species, and so Ockham argued that the various functions performed by the intelligible species can be fulfilled by various dispositions (habitus) of the intellect.[36] Such intellective dispositions are themselves the result of prior causal interaction with the world; Ockham describes their formation through acts of intuitive cognition in the sensitive and intellective souls.[37] Ockham endorses the general claim that "given a sufficient agent and patient in proximity, the effect can be posited without anything else" (William of Ockham, Reportatio II q. 15, prima conclusio). Applied to ordinary cases of cognition, the "agent" is identified as the external object and the intellective disposition, as material and immaterial partial co-causes, and the "patient" is the intellect; the effect is the occurrent act of understanding. For the formation of the intellective disposition, the "agent" is the external object and the "patient" the sensitive and intellective souls. Hence Ockham simply declares it to be the nature of the sensitive and intellective souls that an object is both sensed and understood when it is present or "in proximity." No transduction takes place; sensing and understanding are distinct effects of the same cause, the former proximate and the latter remote.

The causal analysis proposed by Ockham had previously been rejected by other philosophers, such as Durand of St.-Pourçain, for the reason that material objects could not exert a causal influence on an immaterial intellect; Ockham brushes aside this objection by asserting that they can have such an influence (Durand of St.-Pourçain, Quaestio de natura cognitionis, calls the suggestion "absurd"; Ockham's rather brusque retort is in Reportatio II q. 15). Yet in order for an object to have such an influence, the intellective soul must have a potency for conceptualizing either the very object or the very kind of object. Hence the intellective soul is either predisposed to do so, presumably by God's ordering of things, or acquires the power on the occasion of causal contact, presumably by God's intervention. And, indeed, Ockham explicitly admits that God is a partial cause of every act of understanding, at least through establishing the general causal order.[38] It is God, and God alone, who allows such material objects to have causal efficacy on immaterial intellective souls.

Therefore, Ockham offers a "naturalized" version of illumination: the intellect is predisposed to ideate in determinate ways in the presence of different kinds of objects, without the additional (and mistaken) claim that illuminative transduction takes place. Other Scholastics were more forth-coming about this conclusion. Durand of St.-Pourçain straightforwardly held that sensing and understanding are "immanent acts" of the soul, sustained by the ordained causal order, while Peter John Olivi—though giving lip service to illumination—explicitly stated that acts of sensing and understanding occur only by a kind of coordination or harmony (*colligan-tia*) of the faculties of the soul with external events.[39] Much later, Fran-cisco Suárez, in his questions on the *De anima*, would reject both abstraction and illumination, offering instead a version of Olivi's theory, based on a harmony between objects, sense, and understanding.

The difficulty with rejecting transductive mechanisms is stated simply: All of the philosophical problems which drove philosophers originally to postulate transductive mechanisms are left in place, and such problems are not resolved or easily dismissed by maintaining that no such mechanism is called for.

Three separate strands emerge from the rejection of transductive mecha-nisms. The first is that any link between the cognitive faculties of sense and intellect is given up in favor of parallel processes of actualization in each faculty which are ultimately coordinated by God. The second is that the intellect has recourse to a set of predetermined concepts not "derived" from sense, although sensing may be a *sine qua non* condition of their deployment, and these predetermined concepts are construed as dispositions.

The third strand is more subtle. Theories of abstraction and illumination began with a well-defined analysis of the operations of the sensitive soul and attempted to situate a transductive mechanism in their proposed analyses of the intellective soul. Philosophers who rejected the need for a transductive mechanism changed the conditions of the analysis. On the one hand, the contents of the intellective soul no longer had to differ intrinsically from their counterparts in the sensitive soul, whether by being more abstract and universal, or by subsuming sensible species under an exemplar. Rather, inherence in the intellective soul alone was sufficient to be counted conceptual or intelligible—a development which was fostered by the increasing concern with intellectual grasp of individuals. The dif-ference between the sensitive and intellective souls was itself primary, and hence was of itself unexplained, that is, a matter of the basic ontological gap between the material and immaterial. On the other hand, many of the principles which governed the intellective soul also applied to the sensitive soul, and the collapse of the one led to the collapse of the other. Ockham, Durand, and Olivi all sharply modified the Aristotelian analysis of the sensi-tive soul, with Durand and Olivi maintaining that sensing, like under-standing, is an "immanent act" of the sensitive soul, and no longer tied to the physical configuration of bodily organs as functional states. Olivi went so far as to postulate a kind of "spiritual matter" out of which the intellective and the sensitive souls were each composed.

The rejection of transductive mechanisms in High Scholasticism was reflected in developments in Suárez and Renaissance Scholasticism; the three strands of thought emerging from this rejection would dominate the later Scholastic inquiries into the philosophy of mind.[40] Indeed, many of the programmatic concerns present in the philosophy of mind during the early modern period may be seen as extensions of the later Scholastic inquiries— new attempts to resolve the problems set by the agenda of the older debates.

This is especially apparent in the case of Descartes, whose familiarity with later Scholastic philosophy has been well established. (Descartes was hardly alone in this regard; all of the great figures of the early modern period—Arnauld, Berkeley, Hobbes, Locke, Leibniz, Malebranche, and to a lesser extent Spinoza—were thoroughly grounded in the theories of their Scholastic predecessors. Yet Descartes, as the "father of modern philosophy," in many ways stands closest to mediaeval concerns and problems.) Moreover, Descartes can be fitted squarely into the mediaeval agenda. The innovation which more than any other serves to set Descartes apart from his Scholastic predecessors, namely, the modern notion of "mind," directly addresses the problems which come from the rejection of transduction. More exactly, Descartes's theory of the mind was developed in an attempt to resolve the conflicting tendencies present in the third strand described previously. To see why this should be so, a closer look at Cartesian philosophy of mind is in order.

6. CARTESIAN PHILOSOPHY OF MIND

For the Cartesian, the distinction between the living and the nonliving had nothing to do with "soul," but is merely a distinction among types of bodies: "life" is given a purely mechanical account, a description of a class of functioning machines. Other than the (arbitrary) restriction of life to working machines which are composed of certain materials, namely, animal spirits flowing through nerves, we might with equal propriety talk of living watches and dead watches as we do living sheep and dead sheep (see Descartes's letter to Henry More of 30 July 1640, AT III, 182, for this point, also made, though less clearly, in *Passions of the Soul* I Section 6, AT XI, 330-331). Human bodies are no different from the bodies of sheep in this regard. The association of a human body with a Cartesian soul is not causally responsible for human life; the separation of the soul from the body is not the cause of death, but rather death, understood as the breakdown of the bodily machine, is the cause of the separation of the soul from the body (*Passions of the Soul* I, Section 6, AT XI, 330-331). A Cartesian soul is itself a substance, related to but really distinct from the substance which is its associated bodily machine. The exact character of the relation between these distinct substances is a matter of the details of their interaction, but, before considering this, we need to examine the Cartesian soul itself.

The Cartesian soul is a "thinking thing," a *res cogitans*. According to *Meditations* II, a thinking thing is something that "doubts, understands, affirms, denies, wills, refuses, and also imagines and senses [*imaginans quoque et sentiens*]" (AT VII, 28). Descartes defines "thought" in the appendix to his *Replies to the Second Objections* as "all that of which we are conscious of operating in us, and that is why not only understanding, willing, and imagining but also sensing [*sensuum*] are thoughts" (AT VII, 160). Indeed, Descartes more than once speaks of sensations as "confused" thoughts, as when he states in *Meditations* VI that "all these sensations of hunger, pain, thirst, and so on, nothing other than certain confused modes of thinking" (AT VII, 81)—a hallmark of so-called "rationalism." Therefore, thinking and sensing are treated on a par as phenomena which are equally grounded in the same thing, namely, the Cartesian soul itself, distinguished only by degrees of clarity and distinctness.

This conclusion is not affected by Descartes's distinction of three "grades" of sense in the *Replies to the Sixth Objections* Section 9 (AT VII, 436-438). The first grade, the nerve movements, is "common to us and the brutes" (p. 486), and is solely a matter of the mechanical responses of the bodily machine. The second, which "would pertain to nothing but sense, if we should want to distinguish it carefully from the intellect" (p. 437) is the perception of secondary qualities—at least, these are all he mentions—due to the "union" of mind and body. The third grade, the judgements of size, shape, position, done by an (unconscious) calculation based on "the extension of the color and its boundaries," is "commonly assigned to sense," and so called by Descartes a grade of sensory response, "yet it is clear that it depends on the intellect alone" (pp. 437-38). First-grade sensing is clearly nonmental, and Descartes offers a purely causal analysis of it; the second and third grades are states of the mind, the second grade confused and the third grade distinct.

With the location of second-grade and third-grade sensing on the side of the Cartesian soul, divorced from the physiological sense-organs of the bodily machine, Descartes has created a unitary "inner space": the mind. Its structure is minimal. A distinction of subject and object is possible, but the highly articulated Scholastic framework of distinct faculties is not present; it contains only a self and its "thoughts," confused or otherwise.[41] The defining mark of ideas in the strict sense is their character as representations, where representative character is cashed out as the "objective being" of something in the mind. Whether objective being applies to all or at least to some sensations is not clear from the texts.[42] In general, Descartes admits two modes of awareness: (i) direct nonrepresentational awareness of mental contents, and (ii) indirect representational awareness had by means of (at least some) mental contents. Furthermore, at least some ideas—the ideas of simple natures—are taken to be innate, that is, not originally derived from external causes, a claim which is compatible with holding that external causes are responsible for their occurrent conceptualization. (Descartes occasionally flirts with the notion that all ideas "which do not involve affirmation or negation" are innate [letter to Mersenne of 22 July 1641, AT III, 414].)

The relation between the Cartesian soul and its bodily machine is sometimes said to be a "substantial union," as in the *Replies to the Fourth Objections* (AT VIII, 228), and the Cartesian soul is there even called a "substantial form." This "union," notoriously, is supposed to take place through the unique relation of the mind and the pineal gland. But radically distinct substances cannot be united by terminological tricks, and Descartes eventually gave up trying to make an account involving the pineal gland to work, saying that the relation between soul and body is primitive and unanalyzable (see his letter to Elizabeth of 28 June 1643, AT III, 690). Yet basic facts which are descriptive of the relationship between soul and body could be known, succinctly summarized in Part 1 of the late *Passions of the Soul*. First, the Cartesian soul initiates movement of the body through the relationship. Second, the relationship is responsible for the close parallel between changes in bodily state and occurrent sensations. Third, the relationship is responsible for the parallel between occurrent sensations and the associated ideas.

In summary; Cartesian philosophy of mind endorses a unitary "inner space" in which pains, perceptions, ideas, and truths are the immediate subjects of nonrepresentational awareness; at least some of these elements are themselves representational, where "representation" is analyzed as the presence of what is represented in objective being. The assimilation of sensations—pains and perceptions—to ideas and truths is motivated by construing the living body as a well-functioning automaton; the distinction among items in inner space seems to be grounded on the distinction between degrees of clarity and distinctness (although there may be nontrivial distinctions on the basis of representative character); at least some ideas are innate. The "union" of Cartesian soul and bodily machine, while primitive and unanalyzable, is seen in the tight fit between occurrent events in inner and outer space.

The innovation which sets Cartesianism apart from Scholasticism is the creation of the mind, a bold combination of the conflicting tendencies present in the third strand following upon the rejection of transductive mechanisms. Descartes adopted the insight that there was a primitive and irreducible ontological gap, as did his Scholastic predecessors. But Descartes also adopted the Scholastic insight that sensing and understanding should be given a uniform analysis. To embrace both insights, Descartes found it necessary to locate the ontological gap not between the sensitive and intellective souls, as the Scholastics did, but between the bodily machine on the one hand and the Cartesian soul on the other. This move, relocating the gap between the physical and the nonphysical, is at the foundation of the modern notion of the mind: Acts of sensing are thereby classified as nonphysical, and the connecting links to the physiological sense-organs of the body are completely severed.

Descartes's solution, linking sense with thought on the other side of a primitive and inexplicable ontological difference from the body, was understood even at the time as a genuine breakthrough in resolving the problems in the philosophy of mind which had plagued the Scholastics, shattering the Scholastic paradigm.[43] Indeed, the other strands which emerge from the rejection of transductive mechanisms also have a place in Descartes's system. The first strand appears in Descartes as the absence of a link between the physical and the "mental," although events in one sphere are coordinated with events in the other sphere, a coordination ultimately due to God's ordering of the world—a particular instance of a *harmonie préétablie*. As for the second strand, the intellect, or the mind generally, has recourse to a set of "innate ideas" not derived from sense, though sensing is a condition of their deployment. It is no accident that these features, deriving from the mediaeval agenda, also characterize the philosophical views of other "rationalist" thinkers of the early modern period.

Descartes is only one figure in a long chain of thinkers who attempted to resolve the difficulties posed by the rejection of transductive mechanisms. Indeed, it is not unreasonable to see the bulk of modern philosophy of mind as running through an agenda which is essentially mediaeval: whether ideas are acquired or innate, whether abstraction can serve to connect sense and understanding, and, in Kant, the question again posed explicitly—what psychological mechanisms, operating prior to and grounding the possibility of understanding, have to be postulated to account for the facts of mental life? To say this is not to deny the real accomplishments and innovations of the philosophers of the modern period, but to put them in their proper historical and philosophical perspective.

7. CONCLUSION

It is by now a generally accepted thesis in the history and philosophy of science that the creation of the modern "exact" sciences, such as physics and chemistry, is indebted to a long mediaeval tradition; it cannot be understood apart from that tradition; and the eventual failure and collapse of the Aristotelian paradigm was crucial to the formation of modern science. I hope to have suggested a similar pattern for psychology and the philosophy of mind. Yet the Scholastic debates deserve a place of honor in the history of psychology not merely for their historical importance and influence, but because Scholastic philosophy of mind, with its emphasis on a "faculty psychology" and the problem of transduction, may represent a more sophisticated philosophical approach to psychological problems than that found in the early modern period, bearing remarkable similarities to contemporary questions and accounts being developed in cognitive science.

While further research is needed to understand the exact philosophical and historical developments which took place, particularly in Renaissance Scholasticism, the subtlety and penetration of the analyses offered by the Scholastics are unparalleled, and the questions they address can once again be seen as philosophically pressing and acute. It seems apparent, then, that the collapse of a research program may not be the final death, but rather, like the phoenix, it may rise from the ashes at a later date with renewed vigor.

The Ohio State University

Notes

*Versions of this paper were read at the Eastern Division APA Meetings in Boston (December 1986), at the University of Illinois at Chicago (March 1987), at the University of Toronto (November 1987), and at the Conference on Scientific Failure sponsored by the Center for the Philosophy of Science at the University of Pittsburgh (April 1988). In addition to questions and objections from these audiences, I have had the benefit of comments from Annette Baier, Joe Camp, Emily Michael, and Jack Zupko. References to mediaeval primary texts are given in the standard format for the work in question. All translations are mine.

1. If anyone wants to quibble with my use of "mind" as applied to the Scholastics, insisting that this term is misapplied to historical figures prior to Descartes, please feel free to substitute "soul" throughout; no substantive point will be affected. The Aristotelian approach to the mind, whether the discipline be termed "psychology" or "philosophy of mind," was understood to be a scientific enterprise according to the canons of scientific inquiry of the time.

2. Three points of disanalogy should be mentioned at the outset. First, Pylyshyn is concerned with any point in a system in which it is appropriate to describe input as "physical" or "from the environment" and output as "symbolic" or "computational." Therefore, his concerns are more general, not restricted to what in the case of humans we might call the faculties of "sense" and "intellect." Second, Pylyshyn also argues that the transformation of the input to the transducer, described physically, to the ensuing tokened computational event, also described physically, should follow from physical principles. This claim is in the service of his avowed physicalism, and we will ignore it here, since for mediaeval philosophers the intellective soul is paradigmatically non-physical. Third, Pylyshyn takes "meaning" to consist in the rule-governed manipulation of tokens, but this is inessential to his account of transduction; all that is required is that the output be "meaningful" in some sense (which permits Scholastic "mental language" to qualify).

3. The Scholastics vacillate between describing concepts pictorially, with the full vocabulary of "resemblance," "image," and so on, and describing them linguistically, where the concept is the *verbum mentis*; usually both descriptions are present. The theory of mental language, elaborated by Walter Burleigh, William of Ockham, Jean Buridan, and others, explicitly cashes in on the linguistic approach: there is a fully developed grammar and syntax of concepts, which have definitions. Yet even prior to the elaboration of mental language, the notion of the mental "word" is a Scholastic commonplace, found as early as Augustine.

4. The simplified model of Aristotelian cognition presented in Section 3 is deficient in two important respects. First, it only accounts for occurrent sensing and under-

standing, not for cases in which the external object is absent or otherwise causally inactive. To get around this difficulty, two "internal senses" in addition to the common sense are postulated, namely, imagination or "phantasy," and memory. The imagination is a faculty which serves as the storehouse of forms (*thesaurus formarum*) from which such forms could be drawn by memory in the absence of the thing. Second, the model makes no provision for either judgement or discursive reasoning. The former is described in *De anima* III.vi 430a 26-28 as the power of the intellect to engage in "combination and division," that is, to combine (affirm) concepts or to divide (deny) concepts. (No sharp distinction is present in Aristotle between juxtaposition, as in "the white sheep," and predication, as in "the sheep is white"; because of this assimilation the power of combination and division is sometimes taken to reside in the imagination rather than the intellect.) These abilities were taken to presuppose the acquisition of concepts, and therefore to be appropriate to a later stage of analysis.

5. See Aristotle, *De anima* II.ii 413a 22-25: "That which has soul is distinguished from that which does not [have soul] by life; but since 'life' is said in many ways, we say here that a thing is alive if any of the following is present: understanding (νοῦ ς), sensing (αἴσθησις), local movement, as well as the movement implied in nutrition and growth or decay."

6. Aristotle offers a general characterization of "soul": "the first actuality of a natural body structured by organs" [ἐντελέχεια ἡ πρώ τη σώ ματος φυσικοῦ ὀργανικοῦ], (*De anima* II.i 412b 5-6). It is the "first" actuality because there are grades of modal distance from the actual; a sleeping person is in second potency to speech, while someone awake but not speaking is in first potency to speech. The physical body is similar to someone awake but not speaking: It is structured by organs which are connected in the physiologically correct way, but which are not yet animated.

7. In complex entities, such as human beings, are the clusters of principles which individuate the "kinds" of soul distinct entities? That is, how many souls are there in humans—one, two, or three? The answer to this question will turn on views about the unity of substantial form. We need not address this issue here; other animals have only sensitive souls while humans have intellective souls as well as sensitive souls, and any analysis of the sensitive soul in humans must be continuous with the analysis of the sensitive souls of other animals. Hence it does not matter whether the souls are really distinct in humans or not since the principles defining the sensitive soul are the same in humans and other animals, and humans are distinct in kind from other animals—which is a sufficient difference for our purposes here.

8. Aristotle enunciates this principle in *De anima* III.iv 429a 12-15: "Understanding is like sensing, and so it is either a process in which the soul is acted upon by what is understandable—or something else which is analogous to that (ἤ τοιοῦ τον ἕ τερον)."

9. External objects act through the appropriate medium; the details of this causal interaction are dealt with by the appropriate science: in the case of vision, optics. We may ignore the details here, although the tradition of *scientia perspectiva* underlies many of the claims made about perception and the powers of the sensitive soul. Questions arising from this tradition, as for example difficulties dealing with illusion and hallucination—and so the motivation for Peter Aureoli's claims about *esse apparens*—are bypassed in this presentation, as noted in Section 2. A fuller account would take note of these points.

10. Each sense-organ is therefore ensouled, a view known as the doctrine of "animated sense" (note the suggestion that phenomenal appearances are physically interpreted). This doctrine marks one of the great differences between Aristotelian and Cartesian philosophy of mind: For Descartes, sensations are purely mental events, fortuitously corresponding to events in the soul's associated bodily machine; for Aristotle, sensings are largely physical matters, or at least the functional states corresponding to the physical configuration of the organs—see further in Section 6.

11. See Aristotle, *De anima* 412a 6-9: "There is no more need to ask whether body and soul are one than [to ask] whether the wax and the impression it receives [from a signet-ring] are one, or generally whether the matter of each thing is the same as that of which it is the matter; while 'one' and 'being' are said in many ways, the main way is as 'actuality.'"

12. The form of the visible object is present only as a determinate configuration of the eye, which already has a physical form (retina, cornea, and such) animated by the sense-faculty; the identity referred to here is the identity of the form of the visible object with a determinate physical configuration of the animated eye—a pattern of rod-and-cone firings. Furthermore, the identity must hold between ordered sequences of rod-and-cone firings, not a single static pattern, since objects present different color-expanses at different angles and different distances.

The form in the soul and the form in the object have different subjects of inherence. It is a further question whether the forms in themselves, considered without regard to their subjects, are identical or not. However this be resolved, the "identity" of the form in the object and the form in the soul, despite the different mode of inherence, is a matter of *encoding*. Encoding is neither a matter of representation nor of isomorphism. Encoding is not representational since it is the very individual itself which may be encoded. For example, in speaking into a telephone, the actual utterance-token is encoded into a pattern of electrical signals. Encoding need not be an isomorphism since the "code" need not reflect all the features of the "message" in its previous environment (for example, the size of the inscription is not preserved in an utterance of the written message). The point is that in different realms, such as writing, speaking, telephoning, and so on, one and the same pattern, or "form," may be encoded; the given realm determines the exact encoding and embodiment of the form, and the form is identically the same in each of these cases. The mediaeval analysis talks of the "similarity" of the form in the soul and in the object, but this need not have anything to do with resemblance; there were three categorical senses of "sameness" in mediaeval philosophy, namely, identity (sameness in substance), equality (sameness in quantity), and similarity (sameness in quality). To say that the form in the object and in the soul are "similar" is to assert what we would call their identity. At least, such is often the intention; due to Aristotle's comments about "natural similarity" (*De interpretatione* i 16a 7-8) and reasoning based on mental images, some mediaeval philosophers explicitly adopted a resemblance-theory of concepts. But there is nothing in the Aristotelian theory which forces one to hold this.

13. Technically, each sense-modality has a domain of "proper sensibles," as the faculty of vision is associated with the visible (or, more exactly, with color), and may also be able to incidentally discern "common sensibles" (motion, rest, shape, magnitude, number, and unity). Thus the faculty of vision reports on discrete three-dimensional color expanses. In *De anima* III.i 425a 14-28, Aristotle describes the common sense, pointing out that the common sense unifies the proper sensibles of each modality with the common sensibles to produce a distinct sensing of an object. Aristotle also claims that the common sense allows the sense-modalities to be distinguished from one another, but he is not clear that this is by means of the common sensibles or in another manner.

14. The form, viewed solely as the determinate physical configuration of the sense-organ is called the *species impressa*. The form viewed as the determinate actualizing of the potencies which are the sense-faculty is called the *species expressa*. A similar distinction applies to the form as affecting the common sense; the determinate actualizing of the potencies which define the common sense is also called the *species sensibilis*. When this determinate actualization is stored in memory, or at least retrieved by memory, it is called the *phantasm*. The terminology here is not stable. Different authors will regiment the terminology along different lines, and, indeed, individual authors are not always consistent. But this description seems to fit the usage of the majority.

15. The sense-faculty is not totally passive; it is the potency of a living sense-organ, and as such is one step removed from an inanimate receptacle such as a mirror or lump of wax. The point is that sensation must involve an *act* of the sense-organ, which is something an inanimate object could never provide.

16. The argument is simple: Unless there were an agent cause for the actualization of the potency, there would be no more reason for the potency to be actualized at one time rather than another; hence the process would either always be actualized or never be actualized at all, each of which is evidently contrary to experience. Note that this argument does not require an *external* cause—Aquinas accepts this stronger claim, but Scotus rejects it.

17. This description suggests that conformality also provides an explanation of intentionality. If so, then is sensing, which is also explained through conformality, intentional? Some distinctions have to be drawn to avoid conflating the phenomenality of pains, the "pseudointentionality" of sensing, and the genuine intentionality of thinking; drawing these distinctions was an important project in Scholastic philosophy of mind.

18. See Duns Scotus, *Quaestiones quodlibetales* 15.6; there is a similar argument offered in *Ordinatio* I d.3 pars tertia q. 2 n. 486. The "active principle" Scotus refers to here is not directly identified with the agent intellect; rather, Scotus argues at length that the agent intellect and the intelligible species function as partial cocauses of understanding, operating together as an integrated principle (in 15.19-35 and nn. 486-503, respectively). Given that Scotus allows for the possibility that the possible intellect is a partial cause of understanding (see n. 19), his argument is only effective in establishing that there is an active principle which produces the intelligible species— which is all Aquinas's argument, presented in the next sentence, appeals to. The claim that understanding involves an "absolute form" is defended by Scotus in *Quaestiones quodlibetales* q. 13 art. 1.

19. Scotus is less sure about this. In his *Quaestiones quodlibetales* q. 15, he describes two theories, argues for and against each of them, and ultimately does not decide between them—an extremely unusual fact for a mediaeval philosopher! The first theory he describes is like that of Aquinas, where the agent intellect has two distinct functions; the second theory restricts the agent intellect to abstracting the intelligible species from the phantasm, and takes the possible intellect to reduce itself from potency to act. (Such a case is possible for Scotus since he has a complex theory of how things can be "self-movers": see his *Quaestiones subtilissimae super Metaphysicorum libros Aristotelis* IX qq. 17-18; the essentials of the doctrine are alluded to in *Quaestiones quodlibetales* 15.85.)

20. That is, *cum universale ut universale nihil sit in exsistentia*. In this question, Scotus offers a series of arguments showing that there is an intelligible species; he bases his arguments on the possibility that the universal is understood, explicitly putting aside the question of intellective cognition of singulars as irrelevant for this discussion. Scotus, unlike Aquinas, offers a theory of intuitive and abstractive cognition, but the question the theory addresses, namely, the "existence and presence" of the object, does not affect the analysis he gives of abstractive cognition of the universal—a point Scotus explicitly notes in n. 348.

21. Thomas Aquinas, *Summa theologiae* Ia q. 54 art. 4, q. 79 art. 3-4, q. 84 art. 2 and art. 6, q. 85 art. 1, q. 86 art. 1; *Summa contra gentiles* II.lxxvii, *De spiritualibus creaturis* art. 10 *ad* 4 and *ad* 17; *Quaestiones disputatae de anima* art. 4; *De veritate* q. 10 art. 6 *ad* 2 and *ad* 7; *In De anima* III lect. 8 and lect. 10; *De unitate intellectus* n. 111. Scotus describes the process of generating the universal intelligible species from the particular sensible species or phantasm, as in *Ordinatio* I d.3 pars tertia q. 1 and *Quaestiones quodlibetales* q. 15, but generally does not use the term "abstraction" (although in *Quaestiones quodlibetales* 15.53 he does so). Notice that both Aquinas and

Scotus describe the abstraction as proceeding from the sensible species or phantasm, not from the external object itself.

22. While Scotus is reticent about details, the obvious conjecture is that the agent intellect prescinds from the "haecceity," the individualizing differentia, combined with the common nature in the object. However, it is not clear how to reconcile this suggestion, as well as the several places where Scotus talks about the individual particularity of the sensible species or phantasm, with his argument in *Ordinatio* II. d.3 pars prima q. 1 nn. 20-22 that the object of the senses has a real unity which is less than numerical unity. The case of Aquinas is even more difficult: While holding that matter is responsible for individuation, he seems to have changed his mind about whether designated or undesignated matter is the principle of individuation—and, in any case, since the senses take on the form of the material object without its matter, there is a problem in individualizing the sensible species. (Aquinas's offhand remark in *Summa theologiae* Ia q. 75 art. 6 that the senses operate *sub hic et nunc* suggests a possible way out: The individualization accomplished by material conditions combined with the form in the external thing might correspond to the individualizing conditions of here-and-now combined with the form in the sensing). In any event, my discussion does not turn on the precise details of the account of individuation.

23. Universality can be distinguished from commonness, as it is by Scotus (e.g., *Ordinatio* II d.3 pars prima q. 6), but nothing rides on this technical point: The form in the individual must be "general." For Scotus, the common nature is combined with the haecceity in the individual; the common nature is only modally distinct from the haecceity, and so in itself possesses commonness. For Aquinas, there is only a distinction of reason between the form in the object and the form conceived without precision; the form itself includes nondesignated matter, but in the object it is combined with designated matter—according to Aquinas's early doctrines. The individualized form must still be in itself universal; both Scotus and Aquinas offer metaphysical explanations for how this is possible.

24. It is worth emphasizing that this conclusion *only* follows given the premise that the form present in the object is combined with individualizing conditions to become individual—and I am indebted to Walter Edelberg for pointing this out to me. Mediaeval philosophers who were committed realists, such as John of Jandun or Boethius of Dacia, could avoid this conclusion by holding that (i) an individual is composed of forms, at least some of which are universal; (ii) the qualitative difference between universal and nonuniversal features of the individual acts as a "presorting" mechanism, prior to sense; (iii) no transductive mechanism is required. The qualitative difference between the cognitive faculties of sense and intellect, characterized by particularity and universality respectively, is directly traceable to the qualitative ontological differences among the features which compose the individual. Upon coming into contact with an individual, the senses absorb its particular features and "pass along" its universal features, without operating on them in any way, to the intellect, which simply receives them. On this account, the intellect is solely a passive faculty, and no transductive mechanism is required.

There were two large minority traditions during the period of High Scholasticism which endorsed (i)—(iii): the so-called "Latin Averroists" and the "speculative grammarians" (not always sharply distinguished). We will discuss the rejection of transductive mechanisms in Section 5, but for now it suffices to note that two reasons militated against this solution: First, metaphysical realism about universals was in general thought to be too high a price to pay; second, the passivity of the intellect was taken to be contrary to experience.

25. Following Aristotle, Aquinas says that the sheep possesses a "natural estimative power," which is the correlate to human "cogitative power" (*Summa theologiae* Ia q. 78 art. 4 and also q. 81 art. 3; *Summa contra gentiles* II.lx.1; *Sent.* II d.20 q. 2 art. 2 *ad*

5), and he even speaks of the sheep *judging* that the wolf is inimical (e.g., *Summa theologiae* Ia q. 83 art. 1), although the judgement is stigmatized as "unfree." But in these passages it is clear that Aquinas is referring to the sheep's hardware configuration, such that when the sheep's common sense is put into a certain determinate class of physical configurations (including the configuration *wolf*) it will causally actualize the organs corresponding to motivation (e.g., heart, adrenal glands, and so on), producing fear and so triggering avoidance behavior. Different animals will have different hardware links to their motivational organs: sheep flee everything, wolves pursue sheep and flee lions, lions pursue everything. There need be nothing "conceptual" in all this.

26. This is a general difficulty with any property not strictly composed of proper or common sensibles, but most evident for dispositional or modal properties, such as irrationality and rationality. The difficulty prompted later philosophers, such as Ockham and Buridan, to distinguish sharply between the "nominal" and the "real" essence of things, holding that sense merely provides us with a handy grasp on the natural kind (the nominal essence), while it is the task of careful scientific investigation to determine the true nature of the kind (the real essence). This need not entail the rejection of abstractive transduction—the nominal essence could be the product of abstraction—but the argument in the preceding paragraph, that this requires conceptual abilities present in sense, applies as much to nominal as to real essences.

27. Two distinct and independent reasons, not always distinguished, were combined in arguments for this common thread uniting illumination-theories: (i) the claim that unaided human powers are not sufficiently "powerful" to attain conceptual knowledge; and (ii) the claim that it would be impossible to have a local transductive mechanism since transduction requires access to Divine Ideas which no natural power can attain. Whereas (ii) entails (i), the converse does not hold; my discussion is concerned with (ii) since the factual psychological incapacity asserted in (i) leaves the central question unanswered—assuming human cognitive capacities were more powerful, how then *would* transduction take place? (I am indebted to Joe Camp for pointing out this ambiguity in my account.)

28. For abstraction in Bonaventure, see, e.g., *Sententiae* II d.17 q. 1 art. 2 *ad* 4 and d.39 q. 1 art. 2; *Itinerarium mentis ad Deum* ii.6. These passages are typical of many. The discussion of illumination here is taken largely from Bonaventure's *Quaestio disputata de cognitionis humanae suprema ratione*. It should be noted that Bonaventure's primary concern is to safeguard necessary knowledge, which requires illumination as well as abstraction.

29. The sensible species or the phantasm, even taken together with the agent intellect, are not sufficient to produce the intelligible species: see Matthew of Acquasparta, *Quaestiones disputatae de cognitione* q. 2 *ad* 1 and *ad* 12; *Quaestiones De anima* XIII q. 5. It should be noted that Matthew retains the term "abstraction" to describe the production of the intelligible species by the agent intellect and the exemplar, but the terminology is systematically gutted of its customary meaning—just as Aquinas and Scotus use the vocabulary of "illumination" without being committed to any of the theory behind it.

30. See Matthew of Acquasparta, *Quaestiones disputatae de cognitione* q. 2: "the material cause of understanding is the external object, from which the [sensible] species of what is to be known is provided, but the formal cause is partially from within, i.e., from the light of reason, and partially from above." The "light of reason" is the agent intellect, and the exemplar is the partial cause "from above." In q. 1 *ad* 22 Matthew also describes these factors as the formal cause of the occurrent understanding.

31. See Henry of Ghent, *Summae quaestionum ordinariarum* art. 1 q. 2, which is modified and amplified in art. 58 q. 2; *Quodlibeta* VIII q. 12 and IX q. 15. Henry called the process of rendering the clear and lively sensible species into the vague and wispy universal phantasm "abstraction." Henry's theories underwent a marked evolution

during the course of his career; my account is largely drawn from his later writings—in particular, those works composed after 1279, when he rejected the intelligible species. See Marrone (1985) for an excellent discussion of the complexities and subtleties of Henry's theories and development.

32. This claim needs to be qualified. Henry's philosophical development tends toward this final simplification, but it is hardly as direct as suggested here. Even in its mature phase, as represented by *Quodlibeta* IX q. 15, Henry distinguished the possible intellect as material (receptive of the exemplar) and the possible intellect as speculative (able to reflect on its actualization and so gain deeper insight into the exemplar). See Marrone (1985), 136-137.

33. See Henry of Ghent, *Quodlibeta* IX q. 15: God is the *intellegere abditum*. Note that a case could be made for "deferred transduction": with regard to our faculty of understanding, God's agency is just a primitive and unanalyzable function, which suffices for the purposes of a (local) transductive account of understanding. Different problems entirely are involved in accounting for God's direct, and indeed nonconceptual, knowledge of the world, which may be deferred to different investigations—theological in nature. Yet this is to say that there is no general account of transduction, merely a handy local explanation ultimately resting on the inexplicable. While such a conclusion may be theologically sound, it is philosophically unacceptable.

34. The term "special illumination" is also applied to theories in which general illumination is presupposed but God's special and direct intervention is required for certain kinds of understanding, e.g., "scientific" understanding. These theories may be treated without loss of generality as variant forms of general illumination.

35. See William of Ockham, e.g., *Ordinatio* I d.3 q. 6: "[T]he agent intellect is distinguished from the possible intellect in no way; the same intellect has different denominations." See also Jean Buridan, *Quaestiones in De anima* III q. 7, who asserts that the intellect is a simple substance called "agent" or "possible" with regard to different *rationes*. (Buridan, however, retains the intelligible species: see *Quaestiones in De anima* III q. 8.)

36. Ockham argues against the intelligible species at length in his *Reportatio* II q. 15. In his *Expositiones*, he refers twice to the eliminability of the intelligible species (once while discussing Porphyry's *Isagoge* c.ii and once while discussing Aristotle's *De interpretatione*, prohemium). It is mentioned as well in his *Ordinatio* I d.2 q. 8 and d.27 q. 2. He recites the standard list of functions performed by the intelligible species in *Reportatio* II q. 15: to inform the intellect, to unite the object with the potency, to determine the potency to the kind of act, to cause the act of understanding, to represent the object, and to account for the unity of mover and moved. Each function is taken up and discussed, with Ockham arguing that the function is unnecessary or can be accomplished by the disposition. It should be noted that Ockham rejected the sensible species as well as the intelligible species; in this he was certainly preceded by Durand of St.-Pourçain (see Durand's *Quaestio de natura cognitionis, Sententiae* II d.3 q. 6; further evidence that this is Durand's view can be seen in Walter Chatton, *Reportatio* II d.4 q. 1) and Gerard of Bologna, neither of whom, however, posited dispositions.

37. The story Ockham proposes in q. 1 of the prologue to his *Ordinatio, Reportatio* II q. 15, and *Summa logicae* III-2 c.29 is roughly as follows: Beginning with an intuitive cognition in the sensitive soul of an individual material substance or quality, this cognition together with the object "naturally causes" an intellectual intuitive cognition of the same object; given other intellectual intuitive cognitions of the same kind of object (or perhaps of the same object twice), the intellect compares them and constructs an abstractive general concept based on their global similarities and differences.

38. In *Reportatio* II q. 25, Ockham insists that God is an immediate partial cocause of every act of understanding in virtue of sustaining the ordained causal nexus. Because there are no real generalities in the world, Ockham has a difficult time making out the

line that potencies are for *kinds* of responses rather than individualized. But it does not affect the argument if we admit generalized potencies; the point remains that the intellect is predisposed to respond to classes of objects in determinate ways, even if the object itself determines the precise response.

39. See Durand of St.-Pourçain, *Quaestio de natura cognitionis*; Peter John Olivi, *Quaestiones in secundum librum Sententiarum* q. 58 and q. 74. Olivi says that he believes in illumination because "the most distinguished men" hold it, but he adds "I leave the explanation of the difficulties [with illumination] noted above to their wisdom" (*Quaestiones de Deo cognoscendo*, appended to the aforementioned edition). I owe this point to Paul Spade.

40. Oddly enough, the three positions on transduction developed under High Scholasticism remain essentially unchanged in later Scholastic inquiries, despite the greater focus on the philosophy of mind in the Renaissance. Pietro Pompanazzi and Giacomo Zabarella largely repeat Aquinas's account of abstraction; Nicoletto Vernia offers a standard account of illumination; Alessandro Achillini follows Ockham's rejection of transductive mechanisms. Suárez's rejection of transductive mechanisms is noteworthy in part because he develops an original critique of the role of the agent intellect, based on the Aristotelian analysis of causation, showing that transduction cannot take place through any of the Aristotelian four causes.

41. As Descartes says in *Meditations* VI (AT VII, 86), "nor can the faculties of willing, sensing, understanding, *etc.* be called [the mind's] parts, since it is one and the same mind that wills, that senses, and that understands." The unity and indivisibility of the mind is the key factor in the unitary nature of inner space. This is not to deny, of course, the difference between the 'faculties' of intellect and will.

42. Third-grade sensing, which involves judgements dealing with size, shape, and position, certainly seems to allow of objective being. Second-grade sensing is another matter entirely. My claim that representative character is cashed out as objective being may need to be qualified: This certainly holds for "true" ideas, distinct ones, but very confused ideas with virtually no objective reality also present themselves as if they represented something real. Fortunately, nothing hangs on this point, and I leave the matter to Cartesian scholarship to decide.

43. This is not to say that Descartes had no difficulties in spelling out the relation between sensing and understanding, in particular between second-grade sensing and third-grade sensing; but the difficulties are of another character and order: explaining the introduction of the intellect in third-grade sensing as affecting only the degree of clarity or distinctness involved. The real puzzle in Descartes along these lines arises in his account of the relation between a purely intellectual understanding of extensive magnitude and a "distinct *imagining*" of extensive magnitude. But reconciling the role of the pure imagination with the mind's knowledge of which mental imaginative constructions "match" given purely intellectual ideas (and hence which sensed shapes and sizes pure mathematics may be applied to) is not a problem which plagued the Scholastics, for better or worse, and is entirely different from accounting for the transductive mechanism (or lack thereof) linking sense and understanding in the first place.

References

Aquinas, Thomas (St.). All references are taken from *S. Thomae Aquinatis Doctoris Angelici. Opera omnia*, ed. Leonine Commission, Typis Polyglottis Vaticanae 1882-.

Aristotle. All references are taken from the critical editions published in the

Oxford Classical Texts series, as follows: *Aristotelis Categoriae et Liber de interpretatione*, ed. L. Minio-Paluello, Oxford: Clarendon 1949; *Aristotelis De anima*, ed. W. D. Ross, Oxford: Clarendon 1956; *Aristotelis Metaphysica*, ed. W. Jaeger, Oxford: Clarendon 1957. References use the standard titles and Bekker numbers.

Aureoli, Peter. All references are taken from *Scriptum super primum Sententiarum*, ed. Eligius M. Buytaert, The Franciscan Institute: St. Bonaventure, New York 1953.

Bonaventure (St.). All references are taken from *S. Bonaventurae opera omnia*, ed. Collegium s. Bonaventurae, Ad Claras Aquas (Quaracchi) 1882-1902; *Tria opuscula: Breviloquium, Itinerarium mentis in Deum, De reductione artium ad theologiam*, ed. Collegium s. Bonaventurae, Ad Claras Aquas (Quaracchi) 1911.

Buridan, Jean. All references are taken from *Quaestiones in De anima secundum tertiam lecturam*, ed. Jack A. Zupko, University Microfilms: Ann Arbor 1989 (unpublished Ph.D. dissertation).

Chatton, Walter. All references are taken from 'Gualteri de Chatton et Guillelmi de Ockham controversia de natura conceptus universalis,' ed. Gedeon Gál, *Franciscan Studies* Vol. XXVII (1967), pp. 191-212.

Descartes, René. All references are taken from *Oeuvres de Descartes*, publiées par Ch. Adam et P. Tannery, Paris: Cerf 1897-1913, as reprinted by Paris: J. Vrin, 1957-. References are abbreviated 'AT.'

Durand of St.-Pourçain. All references are taken from the *Quaestio de natura cognitionis (Sententiae* II d.3 q. 6), ed. J. Koch, *Beiträge zur Geschichte der Philosophie und Theologie des Mittelalters* Bd. XXVI (1929).

Gerard of Bologna. All references are taken from his *Summa*, ed. Paul de Vooght, *Les sources de la doctrine chrétienne*, Bruges: Desclée de Brouwer 1954.

Godfrey of Fontaines. All references are taken from *Les philosophes Belges* tom.III (eds. Maurice de Wulf and J. Hoffmans) and tom.IV (ed. J. Hoffmans), Louvain University Press, 1914-1921.

Henry of Ghent. References to the *Summae quaestionum ordinariarum* are taken from the edition printed at Paris in 1520, as reprinted by the Franciscan Institute, St. Bonaventure, N.Y. 1943; References to the *Quodlibeta* are taken from *Henrici de Gandavo opera omnia* ed. R. Macken, Louvain University Press, 1972-.

Marrone, Steven P. *Truth and Scientific Knowledge in the Thought of Henry of Ghent*, The Medieval Academy of America 1985.

Marston, Roger. All references are to *Quaestiones disputatae*, ed. a pp. Collegii s. Bonaventurae, Ad Claras Aquas (Quaracchi) 1932.

Matthew of Acquasparta. All references are to the *Quaestiones disputatae de fide et cognitione* ed. Collegium s. Bonaventurae, Ad Claras Aquas (Quaracchi) 1957; *Quaestiones disputatae de anima* XIII ed. A.-J. Gondras, *Études de philosophie médiévale* Vol. 50 (1961).

Olivi, Peter John. All references are to *Quaestiones in secundum librum Sententiarum*, ed. B. Jansen (3 vols.), Collegium s. Bonaventurae, Ad Claras Aquas (Quaracchi) 1922-1926. The *Quaestiones de Deo cognoscendo* is appended to the last volume of this edition.

Pylyshyn, Zenon. *Computation and Cognition*, MIT Press: Cambridge, Massachusetts, 1984.

Scotus, John Duns. Where possible, references to Scotus's works are taken from *Iohannis Duns Scoti Doctoris Subtilis et Mariani opera omnia*, ed. P. Carolus Balić *et alii*, Typis Polyglottis Vaticanae 1950-. Vols. I-VII, XVI-XVIII. References to Scotus's *Quaestiones quodlibetales* are taken from *Obras del Doctor Sutil Juan Duns Escoto (edicion bilingüe): Cuestiones Cuodlibetales*, Introducción, resúmenes y versión de Felix Alluntis. Biblioteca de autores cristianos, Madrid 1968. All other references are to *Joannis Duns Scoti Doctoris Subtilis Ordinis Minorum opera omnia*, ed. Luke Wadding, Lyon 1639; republished, with only slight alterations, by L. Vivès, Paris 1891-1895. I follow tradition in referring to Scotus's revised Oxford lectures on Peter Lombard's *Sententiae* as 'Ordinatio' when the text is given in the Vatican edition and 'Opus Oxoniense' when the text is only available in the Wadding-Vivès edition.

Suárez, Francisco. All references are taken from *Commentaria una cum quaestionibus in libros Aristotelis De Anima*, ed. S. Castellote (2 vols.), Madrid 1978-1981.

William of Ockham. All references are taken from *Guillielmi de Ockham opera philosophica et theologica*, ed. Gedeon Gál, Stephen Brown, *et alii*, The Franciscan Institute, St. Bonaventure, New York: 1967-1985.

III. Social Scientific Perspectives

7. CONVERGING FAILURES: SCIENCE POLICY, HISTORIOGRAPHY AND SOCIAL THEORY OF EARLY MOLECULAR BIOLOGY*

Pnina G. Abir-Am

The ways in which scientists conceive of failure, both their own and that of their predecessors, collaborators, competitors, and significant others, remain to be systematically explored, especially from the vantage of integrative interdisciplinary perspectives in philosophy, sociology, policy and history of science. Important inroads have been made to date on the subject of experimental failure, the negotiation of technical and other errors in the laboratory, and the social process of casting failure as an antidote to discovery, by Branigan (1981), Callon (1986), R. Collins (1988), Knorr-Cetina and Cicourel, eds., (1981), and Lynch (1985), among others.

This paper expands the incipient discourse on failure in science studies by exploring converging notions of failure in science policy, historiography, and social theory, all emanating from a pivotal historical case study in early molecular biology. The paper identifies three main types of failure while focusing on the links between them as translations of failure across different levels of scientific practice. *First*, the paper articulates the anatomy of failure in the implementation of research policy, while distinguishing the process of implementation from that of framing, especially since failure at the level of framing has already been dealt with elsewhere (Abir-Am 1982a, 1984). The problem of assessing what constitutes failure in science policy will be addressed in terms of evaluating the sociopolitical process of the solicitation and the processing of scientific advice in interdisciplinary research (Abir-Am 1986, 1988a; Blume and Spaapen 1988; Chubin 1987; Chubin and Jasanoff (eds.), 1985; Cozzens 1987; Fournier, Gingras and Mathurin 1988; Jasanoff 1987; Porter and Rossini 1985).

Second, the paper explores the historiographical implications of failure in the implementation of policy, especially the tendency to recast a failure of policy as a failure of science, thus denying those who failed at the level of policy or rather those whom policy failed, not only historical priority but also epistemological parity with later, similar but more successful, scientific projects. The negation by historiography of "failure," whether real or apparent, as pertaining to the origins of a successful new discipline, will be shown to derive from the social imagery of science as a guarantor of both truth and practical results, and hence from its ongoing necessity to recast its relativist and pluralistic past in the absolutist terms of the dominant paradigm in the present. The retrospective attribution of failure is meant to preempt the possibility of conceptual dissent with the present paradigm, by reconstructing deviations from it in the past as "mere" technical mistakes or failures. Practitioners may succeed or fail, but all presumably shared the same retrospectively superimposed epistemological framework.

In reality, those often described as failures did not fail simply because in their time the standards by which they have subsequently been judged as failure did not exist. Actually, according to the standards prevailing at that time, they may well have been a great success. The built-in historicity of failure emerges as a sophisticated tool of social control for it denies the legitimacy of conceptual dissent. Moreover, this use of failure further deflects attention from the constitutive role of sociopolitical context in the fate of scientific projects by systematically ascribing failures of context to failures of contents. The retrospective attribution of failure to once successful scientific projects reflects a reluctance on the part of apologetic authors to acknowledge the constitutive, yet often contradictory, roles of policy framing *versus* policy implementation in the subsequent success or failure of innovative, especially interdisciplinary, research (see Abir-Am 1982a, 1984, and the four preceding responses to Abir-Am 1982a to which the latter paper served as a reply; Aaserud 1990; Cueto 1990; Kay 1986; Kohler 1991; Zallen 1992, among others, also deal with the problem of sorting out the rationale for failure of other case studies within the Rockefeller Foundation's portfolio of investments in biological projects in the 1930s and later).

Finally, the paper explores whether the notion of failure could be explained metascientifically as an integral feature of a model of scientific action, in which transitions from micro- and mesolevels to the macrolevel or transitions toward the stabilization and reproduction of structures of signification, legitimation and domination constitute the underlying reason for success; while symmetrical or reversed transitions from macro- and mesolevels to the microlevel, that is destabilization or dissolution of structures constitute the underlying reason for failure. Failure thus emerges as the historically contingent outcome of many accumulating steps and processes of simultaneous destabilization of the three basic properties of social action, meaning, power relations and morality (further details of this sociohistorical model are available in Abir-Am 1987, 28-53).

I. Notions of Failure in the
Historiography of Molecular Biology

Speaking of "converging failures" in connection with molecular biology may sound paradoxical; after all, molecular biology has become synonymous with spectacular success. Since the mid-1950s and especially in the

1960s, rapid progress was made in uncovering the macromolecular structures underlying many biological functions and in demonstrating the universality of the molecular mechanisms of respiration, heredity and immunity. (A critical survey of the historiography of molecular biology is available in Abir-Am 1985b, while a comparison of seven Nobelist lives in molecular biology is available in Abir-Am 1991a; the most relevant perceptions by scientists include Cairns, Stent and Watson [eds.], 1966; Cohen 1975, 1984; Dodson, Glusker and Sayre [eds.], 1981; Kendrew 1967, 1970; Lwoff and Ullmann [eds.], 1979; Monod and Borek [eds.], 1971; Perutz 1980; Rich and Davidson [eds.], 1968.) Molecular biology has come to complement, some even said to displace, evolution as the central problem of biology. (This view was advanced by Jacques Monod, 1910-1976, in his inaugural address as the incumbent of a new Chair in Molecular Biology at the College de France in 1967, published as Monod 1969. A complementary outlook on the relationship between "classical" and molecular biology is elaborated on in Dobzhansky 1964; the views of other leading biologists on this relationship are discussed in Abir-Am 1983/4, 1988b.)

Under these circumstances of rapid success in the post-sputnik era, further superimposed on the great social restructuring of the counterculture age in the 1960s, disproportionate attention was accorded to some episodes in the history of molecular biology, episodes which seemed to fit the prevailing values of success in the late 1960s. The popular obsession with a "winner-takes-all mentality," especially in the Unites States, where culturally prominent enterprises, most notably business and sports, have created a strong cult of simplistic, unidimensional, and outcome-only oriented notions of success, has transformed some very problematic episodes in the history of molecular biology into cultural artifacts. Prominent among those episodes was the story of the discovery of the double helix, which has been circulating for the last twenty years as a "best seller" in many languages, eventually acquiring the ultimate accolade as a movie. (See, for example, Abir-Am 1980, 1991a; Crick 1988, chapters 6 and 7; Judson 1979, Part 1; Olby 1974, Part 5; Stent [ed.], 1980; Watson 1968.)

Under these circumstances, earlier conceptions of molecular biology which did not focus on DNA, but which pioneered the core idea of molecular biology, that of explaining key biological functions across the biological realm in terms of universal macromolecular structures, became problematic for they could not be cast as precursors of the DNA centered view of molecular biology. Even pioneering DNA work done by others than the would-be leaders of molecular biology in the post-double helix era was relegated to "prematurity." For example, several authors designated Avery, MacLeod and McCarthy's paper on the DNA nature of transforming or hereditary material in bacteria of 1944 as "premature" in order to excuse the failure of would-be leading molecular biologists to grasp its significance. Retrospective maturity standards were thus invoked to obscure the prevalence of another, now "mistaken," vision among those who would champion DNA ten years later. (See, for example, Stent 1972 and Wyatt 1972; Avery's contribution was eventually contextualized by McCarty 1985 and Abir-Am 1980, among others.)

Along those lines of conflating the history of molecular biology with that of DNA, other projects in molecular biology which preceded its DNA renaissance of the late 1940s and on, came to be viewed as scientific failures. This paper argues instead that early conceptions of molecular biology, especially those associated with research in physico-chemical morphology in the mid- and late-1930s, did not fail because they made "wrong" scientific choices, as suggested by some of the responses to Abir-Am 1982a from the later form of scientific success stories. Rather, their innovative proponents became entangled in a web of science policy failures, not of their own making, yet failures which led to their termination as leaders of active research projects. (For historical details in support of this argument see Abir-Am 1983/4 and 1987). This termination occurred largely because in the pre-1945 era, preceding the ultimate legitimation of interdisciplinary research as the research strategy underlying a most spectacular outcome, the atomic bomb, the inherently interdisciplinary authority of molecular biology could not obtain legitimation within the then prevailing world of disciplinary traditions, hierarchies and boundaries.

II. THE FAILURE TO GENERATE CONSENSUS AMONG THE ROCKEFELLER FOUNDATION'S ADVISORS ON INTERDISCIPLINARY RESEARCH IN PHYSICO-CHEMICAL MORPHOLOGY

"Molecular biology" is a term, or rather metaphor, which first appeared in "program and policy" documents of the Rockefeller Foundation in 1938. (For reference to sources at the Rockefeller Archive Center see Abir-Am 1982, the paper which first raised the question of the problematic connection between the Rockefeller Foundation and the rise of molecular biology; numerous authors have since addressed various aspects of the Rockefeller Foundation's strategy of supporting science, often though not always in connection with the rise of molecular biology, as evident in Kohler 1991, Aaserud 1990, among others.) Until 1959, when the *Journal of Molecular Biology* was established, the term "molecular biology" was primarily used as a classifying designation in science policy agencies, including the Rockefeller Foundation since 1938, the National Science Foundation since 1951, the British Medical Research Council since 1955, and the French Delegation Generale pour la Recherche Scientifique et Technique since 1958. (See details in Abir-Am 1985-1991, 1992b.) Academic programs, departments, research centers, even international laboratories, carrying this name multiplied rapidly in the 1960s, especially the late 1960s (Abir-Am 1992a).

It is, thus, impossible to start any discussion of success or failure in molecular biology without some understanding of its initial and persisting constitutive context of policy. One of the key episodes leading to the Rockefeller Foundation's use of the term "molecular biology" in its Annual Report of 1938 were the Foundation's prolonged negotiations with a British group of avant-garde scientists, the Biotheoretical Gathering (also known as the Theoretical Biology Club), since 1934, to establish an interdisciplinary institute of "mathematico-physico-chemical morphology" in association with Cambridge University. This metaphor, whose length reflected the egalitarian and multidisciplinary social composition and outlook of the Biotheoretical Gathering, was shortened in 1938 to molecular biology in a dual attempt to both generalize the then emerging policy emphasis of the foundation on experimental biology beyond the soon-to-be-phased-out British enterprise and to capture the then emerging international discourse on the problem of protein structure as the core of the Foundation's interdisciplinary policy of technology transfer from the exact sciences to biology (Abir-Am 1983/4, 1987).

Early in 1933, the Rockefeller Foundation approved a new policy toward natural sciences, with a special emphasis on technology transfer from the exact sciences to "backward, lacking laws and irrational" biology. While the Foundation's policy as a whole could be viewed as a "measure of social reform designed to preempt a radical or socialist challenge" (Offer 1989, 394), the emphasis on biology as the key to social progress was part of the programmatic view of the time, as evident in numerous scientific addresses by science spokespersons and popular authors (see details in Abir-Am 1982a). Nevertheless, the Foundation's new policy had an intrinsic potential for innovation. That potential stemmed neither from its primary, narrowly defined, and positivistic emphasis on technology transfer from classical physical sciences as an automatic guarantor of progress, nor from its condescending perception of biology as an underdeveloped discipline awaiting colonization by the more "progressive," yet, then, theoretically just decapitated classical physical sciences. Rather, this policy's innovative potential stemmed from its derivative focus on interdisciplinary contacts. It could generate a great deal of novelty simply by encouraging contact among otherwise separated knowledge domains, whose boundaries had long been plagued by epistemological qua political conflicts, most notably the threat of reductionism addressed at biology by physics and chemistry. (On philosophical and political aspects of reductionism in biology, see Kitcher 1984 and Abir-Am 1985a.)

However, this built-in innovative potential could be easily eroded if those who implemented the policy, especially the Foundation's officers who mediated between the trustees (the official policy makers), and the scientist grantees whose projects were the target of Foundation influence, were slow to grasp the intrinsically (micro)-political dimension of interdisciplinarity. Essentially, interdisciplinarity constituted a problem of legitimation of a new form of scientific authority, a form in which two or more disciplines provided equally valid and indispensable standards of evaluation. Problems could arise because some disciplines had compartmentalized contradictory explanatory goals and values while being further accustomed to put their own disciplinary ideals above those of other disciplines, which were often viewed as less basic or as objects of reductionism. Such views prevailed at the boundary between the physical and the biological sciences, which also signalled the metaphysically charged boundary between the living and nonliving worlds (for details on this debate in the late 1920s see Abir-Am 1991b). Thus, the possibility of generic conflict over boundary definitions and disciplinary priorities was built into the policy definition from the beginning, yet no specific guidelines were developed to address it.

A major source of both enabling and constraining factors in the implementation of a policy with emphasis on interdisciplinary research, such as the one initiated by the Rockefeller Foundation, pertained to its resorting to advisory systems. On the one hand, innovative or nonconventional research could be promoted by seeking advice from scientists who pioneered themselves interdisciplinary explorations. On the other hand, since most scientists, especially the established ones who were likely to be approached as advisors, often made their scientific reputations in terms of contributions to a given discipline's mainstream problems, their view of new, interdisciplinary research would largely depend on their subsequent exposure to such research; if such exposure was minimal, as could often be the case with busy, established science administrators, their own history of participation in conflicts over disciplinary boundaries, or their perception of nonscientific aspects of the proposed projects would take precedence in forming their judgement.

The contradiction at the core of the Rockefeller Foundation's policy revolved around its ongoing search for legitimation of interdisciplinary research through consensus of advisors, some of whom were being solicited because of their disciplinary expertise or sheer accessibility as current or former grantees, rather than their interdisciplinary background. As no special guidelines were developed to select advisors while minimizing disciplinary and other sources of conflicts of interest, or to process their advice in ways which discounted irrelevant or biased advice, the officers became captive of a futile quest for an implausible consensus as the supposedly only valid basis for decision making. This failure to admit the political grounding of disciplinary consensus in science and the vulnerability of new, interdisciplinary projects, led to the implementation of a policy of "unintended consequences," namely, of terminating rather than consolidating the interdisciplinary projects that had initially been targeted for encouragement.

The Rockefeller Foundation's officers ran into the problem of assessing advice on a large scale when they sought to consolidate the support of a collaborative team of grantees in physico-chemical morphology at Cambridge University, with a long-term grant. This procedure of consolidation usually followed officer satisfaction from the grantees' research under short-term support. It aimed to give the Foundation an ongoing opportunity to strengthen the sort of research it included in its policy while further consolidating the position of its favored grantees in their home institutions, usually by requesting that those home institutions accept future responsibility for investments initiated by the Foundation (for details see Abir-Am 1982, Aaserud 1990, Kohler 1991).

Since 1934, the officers tracked the Cambridge team as prospective grantees through their first comprehensive paper on physico-chemical morphology, in the *Proceedings of the Royal Society* (J. Needham, Waddington and D. M. Needham 1934). The paper embodied both a topic and an approach which the Rockefeller Foundation sought to promote: the application of physico-chemical techniques to biology through the collaboration of scientists from diverse disciplines. Moreover, the paper was communicated by the Royal Society President, Sir Frederic Gowland Hopkins, whose programmatic speeches on the relevance of biology for social progress were a main inspiration behind the Foundation's new policy, and were widely quoted in its "program and policy" documents justifying the new policy (Abir-Am 1982a). The three authors of the papers were the biochemists Joseph Needham and Dorothy Moyle Needham, and the experimental embryologist Conrad Waddington. On advice from Hopkins, who was professor of Biochemistry at Cambridge and head of its Dunn Institute, the officer in charge of European projects, Wilbur E. Tisdale, entered into contact with Joseph Needham, who had recently been elected reader in biochemistry at Cambridge University (*ibid.*).

Between 1934 and 1937 the Needhams and Waddington received several grants, from the Rockefeller Foundation, usually on an annual basis. The Foundation, represented by its European officer Tisdale and by the Director of the Natural Sciences Division, Warren Weaver, expressed its interest in consolidating the research in physico-chemical morphology in various meetings with J. Needham and Hopkins, during which a variety of schemes were entertained. Prominent among those schemes was an interdisciplinary research institute of mathematico-physico-chemical morphology, which was expected to incorporate not only J. Needham and Waddington but also a score of biologist, physicist and mathematician collaborators, including Joseph Henri Woodger, John Desmond Bernal and Dorothy Wrinch, as well as additional supportive staff. They called themselves the Biotheoretical Gathering (better known in the secondary literature as the Theoretical Biology Club) and included a dozen men and women scientists and philosophers, mostly based in the Cambridge-London-Oxford triangle, all intent on revolutionizing biology along the lines of the theoretical revolution in physics of the 1920s. Their 1935 proposal constitutes the single most elaborate exposition of interdisciplinary ideas and collaborative projects in early molecular biology (for archival evidence see Abir-Am 1983/4, 1987).

However, by late 1936 it became clear to Needham and Waddington that their various proposals could not be implemented because of their precarious position at Cambridge University. Large scale or interdepartmental investments by the Rockefeller Foundation required both solicitation and some matching contribution by the institution of the respective grantees. Professor Hopkins, the grantees' administrative superior, did not command sufficient support within the university to secure a formal request from the Rockefeller Foundation. Under these and related circumstances of tension with the conservative university authorities, Needham and Waddington reached the conclusion that their best chance to obtain long-term support from the Rockefeller Foundation revolved around transferring their collaboration to a private research institute on the outskirts of Cambridge, the Strangeways Research Laboratory, whose director Dr. Honor Fell was sympathetic to their interdisciplinary and innovative research and agreed to submit their proposal to the Rockefeller Foundation (for details and references to Dr. Fell and the Strangeways Laboratory see Abir-Am 1988a).

In order to transfer the collaborative project in physico-chemical embryology to Strangeways, Tisdale consulted with two of its Trustees, Professors Henry Roy Dean and Sir Henry H. Dale, as well as with Sir Edward Mellanby, the Secretary of the Medical Research Council, a government agency which had been providing some support to the Strangeways prior to Mellanby's assuming the Secretary role in 1933. Dean was at the time incoming Vice Chancellor of Cambridge University while Dale, a 1936 Nobel Laureate, was Vice President of the Royal Society and Director of the National Institutes for Medical Research. Both Dean and Dale expressed lack of confidence in Needham, but they dissociated their dislike of Needham from the projected support for Needham and Waddington's research at Strangeways on the ground that they had full confidence in its Director, Dr. Fell, who had supported the transfer of their research. Mellanby, too, expressed caution about Needham, but said that he supported the transfer of his project with Waddington to the Strangeways (*ibid.*).

These subtleties, grounded in opposition by establishment figures to Needham and Waddington's social and political radicalism or support of leftist causes at Cambridge University, evaporated when Tisdale and Weaver discussed the response of these three science administrators. Though Tisdale understood that their reservations were not grounded in scientific judgement, and that they were unlikely to directly intervene against grantees accepted by Dr. Fell, the reservations revealed by the advisory process, which mixed politics with policy and with science, alarmed Weaver who remained especially concerned about the personal opinion of Dale and Mellanby about Needham, even though they themselves dissociated their opinion from advice on the project. Weaver further brought their opinion to the attention of the Rockefeller Foundation's then new president, R. B. Fosdick, who, however, left the decision to the officers.

At this point, Weaver instructed Tisdale to collect the "European opinion" on Needham while he set to collect the American opinion. Each one of them contacted six advisors. Tisdale's advisors, despite Weaver's request for the "European opinion," included only one from outside England, while four of his six advisors were from Needham's own institution, Cambridge University. Out of these six, four were positive and two were negative. The positive advisors included figures from outside Cambridge University, who were also involved in physico-chemical applications to biology, such as Nobel Laureate Otto Warburg of the Cell Biology Institute in Berlin and Ian Morris Heilbron of the Department of Physical Chemistry at the University of Manchester. Two additional advisors from Cambridge University who were also involved in physical and chemical applications to biology, the Professor of Biology, David Keilin, and the neurophysiologist, Edward Douglas Adrian, later a Nobel Laureate, were also positive. However, two additional advisors from Cambridge University, the professor of zoology, James Gray, and the professor of colloid chemistry, Eric Rideal, were negative. All the advice collected by Tisdale was oral, casual and brief. The negative advice was particularly enigmatic since no context was supplied to interpret Gray's and Rideal's apparently personal dislike of Needham. Neither commented on his research project or on the prospects of its transfer to Strangeways. It is not obvious why Tisdale consulted with Gray and Rideal, except for his habit to visit with departmental chairs as a way to collect intelligence on prospective grantees.

In contrast to Tisdale, Weaver sought and obtained mostly written advice while further asking his advisors to evaluate not only Needham's standing as an investigator but also the prospects of the collaboration between Needham and Waddington and its transfer to Strangeways. However, the advisors were given neither a document, such as a proposal, nor any specific guidelines as to whether to focus on the investigators' previous record, their future promise or the necessity of transferring their collaboration to another institution. This vagueness on Weaver's part was further coupled with prejudging intimations that Needham had already been viewed as controversial by some previous advisors (the unspecified allusion was to the reservations of the three science administrators whose considerations were more policy and politics rather than scientifically oriented, a fact which was not lost upon both Tisdale and Weaver).

The "American advice" which reached the Rockefeller Foundation in response to such vague and partial guidelines, though offered mostly in writing, was rather heterogeneous, and hence difficult to compare, let alone quantify. Nevertheless, the "canvass" of the American advice paralleled the

European, or mostly British pattern: Advisors who were involved in physico-chemical applications to biology and thus had direct experience with the prospects and problems of interdisciplinary research tended to be favorable, while those associated with more traditional disciplinary visions, either in zoology or chemistry, were negative.

Advice classified as positive came from Ross Harrison of Yale, Viktor Hamburger of Washington University and Alexis Carrell of the Rockefeller Institute for Medical Research in New York City. Harrison and Hamburger's views were the most relevant among all the advisors, not only because they had direct experience with experimental embryology, and provided well argued and balanced judgements, but especially because they directly addressed the problem of the complexity of standards of evaluation in interdisciplinary research. Harrison was himself involved at the time in a collaboration with the British physicist turned molecular biologist, W. T. Astbury, of the University of Leeds, and argued most eloquently that a great deal of interdisciplinary bridging, via prolonged collaboration among scientists trained in different disciplines, must be accomplished before a new interdisciplinary field can consolidate itself. Despite their detailed, well argued and highly relevant advice, Hamburger's opinion was wrongly classified as negative (Hamburger 1986, 1988), while Harrison's was given the same weight as the casual and biased single phrase given by Gray or Rideal.

The other three American advisors, two zoologists and a protein chemist, were explicitly negative. George Streeter of The Johns Hopkins objected to Needham's physico-chemical approach to embryology on the ground that he was not trained as an embryologist, and hence had no chance to acquire the presumably inborn instinct of an embryologist. Benjamin Willier of Rochester University expressed some plausible criticism, namely, that Needham and Waddington sought to claim a distinct chemical identity for the organizer at a time the problem had not yet been solved, but he failed to reveal that he had a vested interest in harming the prospects of Needham and Waddington since he and Waddington were "scientific outs" or had advanced competing theories. The Rockefeller Foundation officer F. B. Hanson, who served as assistant director to Weaver in the New York office, was aware of Willier's conflict of interest, but failed to take it into account upon classifying and counting Willier's opinion. Negative advice also came from J. Murray Luck of Stanford, who criticized the chemical accuracy of Needham and Waddington's work in biochemical embryology, while overlooking the fact that standards of chemical accuracy, though desirable for a relatively new discipline such as biochemistry still in a process of "proving" its scientific status, were not quite applicable to embryological material of the sort Needham and Waddington had been experimenting on.

The European and American advice was eventually lumped together, classified and quantified into six "pros" and six "cons" by Assistant Director

Hanson, the same person who misinterpreted Hamburger's relevant and positive advice, as negative. Director Weaver, in his turn, was primarily disturbed by the fact that his search for broad consensus, and hence for an unproblematic derivation and ad hoc ratification of a projected policy decision, proved elusive. Though he was aware of the sources of bias which rendered most of the solicited advice irrelevant, and detailed them in a letter to his European associate director, Weaver was reluctant to admit that he failed so sharply with the process of soliciting advice, a process which he considered not only a pillar of his policy but also the guarantor of an objective scientific opinion. He did not wish to assume responsibility for a "real" decision, namely, one which could not be justified as derivative of advisory consensus. Furthermore, Weaver was impressed with Hanson's quantification of the heterogeneous advice which further included misclassifications. Having had a strong belief in mathematics' capacity to order physical reality, and an arithmetical disposition in policy, Weaver finally processed the meaning of six "cons" and six "pros" as a "canceling out" or zero, and hence reached the bizarre conclusion that the scientists' advice should be ignored altogether (see Abir-Am 1988a for archival evidence).

Weaver remained disturbed by the elusiveness of consensus on physico-chemical morphology, even though he well understood the inevitability of bias in evaluating interdisciplinary research, and the personal, as opposed to scientific, nature of most of the negative comments on Needham. As a result of his indiscriminate solicitation of advice, an approach stemming from his increasingly becoming accustomed, as he became more conscious of his managerial power, to treat scientific advice not as genuine input into the decision-making process but as a convenient mechanism for ad hoc ratification of decisions already taken. Hence, the project in physico-chemical morphology came to epitomize not only Weaver's long frustrated hopes for a large and prestigious investment at Cambridge University, but also the contradictions of his entire policy, while highlighting the resistance to interdisciplinary research, and hence to his policy, among influential segments of the scientific community.

Moreover, just at that time, the entire policy of Weaver's division had been strongly criticized in front of the Rockefeller Foundation's then new President, Raymond Fosdick, by Sir Edward Mellanby who further attempted to influence Fosdick to change that policy. Mellanby, who was in charge of British governmental policy for biomedical research, disapproved of the Rockefeller Foundation's policy toward biology which was conceived by physical scientists, such as Weaver and Weaver's mentor, the former Rockefeller Foundation President, Max Mason. Mellanby transmitted his views to Fosdick at a time Weaver sought to involve Fosdick in the decision over the project in physico-chemical morphology. As a result of this additional entanglement of his policy, Weaver may have thought that he needed a "sacrificial lamb," to appease Mellanby's increasingly assertive stance over the role of the Rockefeller Foundation in the direction of the biomedical research in England (for details on the Medical Research Council's policy in inter-war England see Abir-Am 1988a and Austoker and Bryder [eds.], 1989). Indeed, Mellanby's criticism of his entire policy prompted Weaver to request that a committee of appraisal be appointed to evaluate his division's projects in 1938 (for the evaluation of this committee see Abir-Am 1982). Ironically, the coincidence in the timing of both Mellanby and Weaver becoming more aware of their respective powers as scientists turned research administrators may have also contributed to Weaver's decision to ignore the scientific advice while responding to policy considerations only.

One also notes that, on the whole, positive advice came from outstanding and innovative scientists such as Adrian, Carrell, Hamburger, Harrison, Heilbron, Keilin and Warburg, while negative advice came from scientists whose main accomplishments were in the domain of organization and administration, not innovation. The question thus persists whether Weaver could have used this implicit alignment of the advisors as innovators *versus* administrators as an indicator of which advice to follow. It is plausible that scientists who were not themselves innovators but rather perpetrators and administrators of research traditions, a category to which Weaver himself belonged, were more likely to conceive of the temporary power they are granted as advisors as a policing opportunity. In their zeal to reassert order and control, they tended to perceive interdisciplinary endeavors as disorder; after all, interdisciplinary research highlights that disciplinary boundaries are often obstructions to both justice and knowledge, though those same boundaries could be, and often are, very useful for maintaining disciplinary control and order (for such an interpretation see Abir-Am 1982a, 1992b).

In coming to summarize the generic types of failure involved in the Rockefeller Foundation's involvement in this case study of interdisciplinary research in a molecular biological framework in the pre-World War Two period, it is useful to distinguish between failure at the sociopolitical level of solicitation of advice, that is asking the wrong people or not asking the right ones, and failure at the semantic level of inadequately processing the meaning of the obtained advice.

Several types of failure can be observed at the level of solicitation of advice. First, many advisors were selected because of their accessibility to the officers on their route to visit various Foundation investments, rather than in terms of expertise. This also accounts for the unusually large number of advisors from the grantees' own institution, even though the officers were aware that in such cases, internal administrative and personal problems tended to spill over into scientific judgement. In a sense, the problem began when advice was solicited without a proper strategy of selecting advisors according to their capacity to illuminate the decision in question, i.e., the transfer of a long term project to the Strangeways laboratory. In addition, no effort was made to reach expert European researchers, such as Jean Brachet in Belgium and Johannes Holtfreter in Germany. Thus, the first type of failure was a failure to select relevant advisors, rather than easily accessible ones (for a recent history of modern embryology see Gilbert [ed.] 1991).

The second type of failure involved the lack of any concrete guidelines to the advisors as to the unit of their analysis; most of them focused their judgement on the personality of one investigator, or his past record, even though the Foundation's decision pertained to the transfer of a new, collaborative project to another institution. Third, the guidelines that were given to the advisors in the US were confused while prejudging the case since Weaver revealed to them that Needham was viewed by an unspecified some as controversial. This was a failure in both conveying and maintaining the impartiality required for reaching an independent assessment.

The fourth type of failure involved confusion of scientific and policy questions: Advisors were asked not only to evaluate Needham's scientific contributions, but also whether his project with Waddington should be transferred to the Strangeways Research Laboratory. The latter was a policy decision which depended on the then changing power relationship between several policy makers, including the university authorities and individual departments, departments and laboratories with vested interest in the various disciplines involved directly or indirectly in the interdisciplinary project of physico-chemical morphology, and especially the emergence of a new outlook within the Medical Research Council with regard to the Rockefeller Foundation's support of biomedical research in England. Yet, advisors were not given information on the complex policy context of the transfer decision they were asked to comment upon.

A multidimensional failure could also be detected at the level of processing the advice. First, advice was processed as if it were inert and as if its contents were not profoundly constrained by its disciplinary, institutional and cultural context, a context which further included a variety of conflicts of interest. For example, though Tisdale was aware that negative advice from Cambridge figures related to administrative problems they seemed to have had with Needham, and recommended to discard those opinions as matters which did not concern the Foundation, Weaver remained concerned with the opinion of figures of authority, which he seemed too eager to both misunderstand and substitute for more relevant advice.

In addition, minor reservations were misconstrued as negative advice, both in the case of the three British science administrators and in that of Hamburger, a recent refugee from Germany. In both cases the misconstrual may have stemmed from cultural misunderstandings on the part of the American officers who failed to realize that, for Europeans, certain forms of reservations do not constitute negative opinion. In addition, Director

Weaver failed to detect these interpretation problems on the part of both Tisdale and Hanson. Thus, what was classified in many instances as negative advice was not meant to be a negative input into the decision making, but merely an evaluative nuance mediated by cultural distinctions and emphasis. The Rockefeller Foundation officers appeared to apply American cultural standards for judging the standing and opinion of both European grantees and advisors, a procedure which swung the entire process of evaluation in the direction of the entrepreneurially minded grantees. In this manner, the Foundation failed to pursue its own guidelines for pursuing a policy geared to advance interdisciplinary projects, favoring instead empire builders who knew how to manipulate the officers' expectations.

At the same time, there was a failure in developing policy guidelines suitable for interdisciplinary research, for the case study reveals a failure to differentiate between the value of casually given advice of little consequence, and that of well argued, detailed and relevant advice: Each advisor's opinion counted as one unit, regardless of its insight or pertinence. Finally, the procedure of quantifying the advice in equal numerical units prevented any subtle nuances of partial support from emerging. The officers' failure to separate relevant from irrelevant advice eventually led Weaver to ignore the scientific advice altogether, while missing the benefit of that part of the scientific advice which was especially pertinent and insightful, namely, that of Harrison and Hamburger.

The above sequence of accumulating failures in both soliciting and processing scientific advice with regard to interdisciplinary research, produced an outcome of unintended consequences. On the one hand, the Foundation's prolonged negotiations which involved delicate intra-university matters, such as the salary of university officers and the transfer of their activities outside it, brought negative light from the Cambridge University authorities on both the Foundation's officers and their grantees. At the same time, the Foundation's persistent interest and ongoing judgement that its project was very central to the Foundation's strategy, prevented Needham and Waddington from seeking alternative sources of support. Thus, the unexpected circumstances of terminating Foundation support for once a favorite project, especially at a time that interdisciplinary research was favored neither by universities, nor by governmental agencies, thus brought the productive collaboration in physico-chemical morphology to an end. By then, it included about a dozen joint papers, many of which reflected collaboration with half a dozen foreign scientists from all over Europe and America, or precisely the sort of activity which the Foundation's policy sought to stimulate.

During and after the Second World War, Needham and Waddington did not have an opportunity to collaborate on physico-chemical embryology, since both left Cambridge, the former for about a decade, while charged with various cultural missions on behalf of science, first in China and later at UNESCO, while the latter served in operations research during the war and joined the University of Edinburgh in 1947 (Needham 1946, 1976). It was not until the 1960s that their pioneering research on physico-chemical approaches to the organizer problem was revived (Gilbert [ed.] 1991; Horder and others [eds.] 1986; Needham 1968; Waddington 1963, 1969, 1975). Finally, in the late 1980s, their initial emphasis on molecular embryology and topobiology has become, once again, in vogue (Edelman 1988).

The failures of the Rockefeller Foundation in implementing its own policy and the impact of those failures upon terminating the research in physico-chemical morphology had been reconstructed by some scientists and historians of molecular biology as a scientific failure of the research program in physico-chemical morphology (for example, see responses to Abir-Am 1982a). The tacit assumption being that if a research program terminates, then it must be because it ran out of scientific promise. Thus, a project is retrospectively stripped of any promise or novelty it might have had in its time, in the name of later events, whose success becomes a source of absolutist interpretation for both the present and the past.

Thus, it was pointed out, on the basis of later developments in molecular biology, that physico-chemical morphology operated at the more complex, and hence molecularly insoluble level of tissue, or multicellular matrix, as opposed to the unicellular matrix of bacteria and phage, the simpler "organismic" level which brought success to molecular biology. Other arguments stressed the problem arising from the project's discovery that a great many various chemicals and physical stimuli could serve as organizers, thus presumably minimizing the possibility that the organizer's biological activity could be explained in terms of a specific molecular structure (Horder and others [eds.] 1986, Gilbert [ed.] 1991).

These plausible justifications cannot however obscure the fact that the research program in physico-chemical morphology did in fact terminate not as a result of a necessary exhaustion of its scientific promise, or dilution of its original vision in the hands of unsuitable investigators, but rather as a result of a sequence of contingent failures in policy implementation. It was not the lack of an immanent "logic of discovery" as a guarantor of future scientific truths, which this interdisciplinary research project lacked, but rather it "failed" to find a policy context in which transdisciplinary scientific knowledge with its attendant relativism of scientific authority could be legitimized. This was to happen only in the post-World War Two period, especially in the 1960s when indeed molecular biology became not only a success story but a cultural and a policy artifact (Abir-Am 1992a,b).

The question thus persists as to how we can deconstruct this retrospective historiography so that the translation of a policy failure into a scientific one becomes intelligible in metascientific or social-theoretical terms. One way to accomplish an interpretation of the above translation is to

reconceptualize the policy failure as a failure in a transition from micro-
and mesolevels of scientific action to macrolevels of stabilized and reproducible scientific structures. In such a model, the policy failure is viewed
as a failure of stabilizing and institutionalizing emergent, metastable
structures of meaning, social order and power relations. To make sense of
this suggestion, the paper proceeds to outline a model of scientific action
and structure, or in Giddens's terminology for social theory (1979, 1986) a
model of structuration, while further illustrating its working with the
historically authentic case study of the Rockefeller Foundation's support
of physico-chemical morphology in the 1930s.

III. THE MEANING OF FAILURE IN A MODEL OF SCIENTIFIC CHANGE
TRANSITIONS BETWEEN MICRO-, MESO-, AND MACROLEVELS
OF SCIENTIFIC ACTION

If the project in physico-chemical morphology was not a failure in the
scientific sense, at the time of its abrupt termination, then what type of
failure was involved in the phasing out of this research project from the
scientific repertory for an entire generation? We suggest that the failure
in question can be best understood as a failure to complete a transition
from meso- to macrolevel of scientific action. The project's ongoing existence for several years meant that minimal or temporal stability was
accomplished via a successful transition from the microlevel of individual
investigators to the mesolevel of a collaborative team with a distinct social
reality. This relative measure of stability was further reinforced throughout the several years of the project's duration as a highly productive and
continuously funded social endeavor.

Yet, once endowment with long-term support became impossible for the
above mentioned policy reasons, the project failed to accomplish a second
transition, that from its mesolevel or level of intermediary stability to the
macrolevel of "permanent" institutional existence. For institutionalization
was necessary to ensure reproduction of new knowledge claims across time
and space via the stabilization of objectified social structures such as
routinized positions, embeddedness in a broader scientific tradition and
the granting of power in the form of direct control over a certain amount
of human and material resources. Indeed, molecular biology became a
recognized field not when its acclaimed discoveries were made in the 1940s
and 1950s but especially in the 1960s, when institutionalization became
possible on a broad scale (Abir-Am 1992a,b).

In order to clarify the strategic role of the above mentioned transitions
for the stabilization of new metaphors, such as "physico-chemical morphology" or "molecular biology," as objective transdisciplinary structures of
signification, it is necessary to explore a model of scientific change which
combines both process and outcome or structure. Such a model intends to
bridge the prevailing split between positivist macrosociologies (e.g., struc-

turalism and functionalism) which focused on the problem of the reproduction of the social order but neglected its production, and interpretive microsociologies (e.g., interactionism and ethnomethodology) which focused on illuminating the neglected problem of the production of the social order, but gave relatively little attention to its reproduction (R. Collins 1988, Giddens 1979, Knorr Cetina and Cicourel [eds.] 1981).

The dynamics between contingent action and objective structure has been described by Giddens, in his theory of structuration, the object of which is to explain "the conditions which govern the continuity of and dissolution of structures and types of structures" (Giddens 1979, chap. 2). Giddens views the relationship between interaction and structure as being mediated by modalities. Since interaction has three basic features (communication of meaning, of power relations and of morality or normative order), Giddens specifies three modalities, each mediating between the features of interaction mentioned above and the respective structures. Thus, the feature of communication of meaning in interaction is linked by the modality of interpretive schemes to structures of signification; the feature of power relations in interaction is linked by the modality of facilities to structures of domination; and finally, the feature of morality or sanction in interaction is linked by the modality of norms to structures of legitimation. Schematically, these links may be represented as:

Interaction	communication of meaning	power	morality
(modality)	interpretive scheme	facility	norms
Structure	Signification	Domination	Legitimation

Just as communication of meaning, power relations and morality are integral elements of interaction, so signification, domination and legitimation are merely analytically separable properties of structure. Indeed, structures of signification can be (and often are) analyzed as systems of semantic rules; structures of domination are often analyzed as systems of resource distribution; and structures of legitimation are often analyzed as systems of moral rules or values (ibid.).

The explanatory capacity of this triple model pertains to the idea that the application of each modality, though dependent upon the respective structure, reconstitutes that structure at the same time. It is this duality of the structures (of signification, domination and legitimation) as both a medium and an outcome which allows for change to be inherent in any conceptual, social, or political system. There is room for ambiguity in the interpretation of modalities (interpretive schemes, facilities and norms, respectively), and thus there is also room for interpretation, contestation and negotiation (ibid.).

In the case study of physico-chemical morphology, the collaboration between Needham, Waddington and their associates both depended on and reconstituted newly objectified structures of signification (the search for the molecular structure of the organizer as an explanation of the problem of biological organization in precise physico-chemical terms), newly objectified structures of domination (the attempt to acquire scientific authority over the new transdisciplinary domain of physico-chemical morphology and its special resources such as the projected, long-term funding by the Rockefeller Foundation for materials, equipment and research assistance), and newly objectified structures of legitimation (the attempt to gain recognition for physico-chemical morphology from established scientists in the "neighbouring" disciplines of embryology and biochemistry).

The constitution of such objectified structures depended on ongoing processes of objectification or modalities, processes involved in the transformation of subjective meanings into objective facticities (Berger and Luckman 1967; Schutz 1962, 1964, 1967). In the case study of physico-chemical morphology processes of interpretation of various physico-chemical experiments on the organizer transformed Needham and Waddington's subjective interest in a molecular approach to the problem of biological organization into the potential structure of a new field. But in order to complete this transformation, their emphasis on the value of interdisciplinary research had to be legitimized and their resources, both human and material, had to be consolidated. Such processes of objectification, pertaining to the values of transdisciplinarity and of power or command of resources, persisted throughout the project's duration, since each annual renewal reaffirmed the legitimation of transdisciplinarity and the consolidation of its power over new resources. These processes pertained to annual officer consultation and decision to renew, to the grantees' ongoing experimental, theoretical and collaborative activity in physico-chemical morphology, and on the recognition of transdisciplinary research by two figures of authority, the Cambridge professors of biochemistry and biology, Hopkins and Keilin (Abir-Am 1988a).

In this sense, the renewed and ongoing objectification processes precipitated a new, metastable state, as a transitional point between the contextuality of individual action and the objectivity of social structures. The transitional state of the research project in physico-chemical morphology reflected such an intermediacy of stability because it possessed a social reality beyond individual contingency, yet it lacked the capacity of scientific disciplines to reproduce over time. In analogy with metastable states known in physics, chemistry, biology and anthropology (see Prigogine and Stengers 1984 for the natural sciences, and Turner 1964 for anthropology), the social condition of physico-chemical morphology was also short lived and sensitive to disturbances in the global system. Such interferences tend to revert the metastable states to the stable microlevel of individual action or to "advert" them to the stable macrolevel of social structure. The obvious importance of metastable states is to facilitate change by converting occasional fluctuations or contingent social combinations into newly stabilized structures or a new type of order. In the case study of physico-chemical morphology the stabilization of a new, transdisciplinary order in science depended upon ongoing opportunities to consolidate a transient collaboration.

Though the transition from the microlevel of individual action to the metastable level of transitory groups is a necessary step toward scientific change, it is not a sufficient step. Only the second transition, from the metastable state to the macrolevel of enduring scientific institutions can transform the ever present possibility of change into an actuality. However, the second transition can only occur under historical circumstances favoring the solidification of local, precarious, new metastable structures into well bound institutions or macrostructures. The failure to collect and process advice properly led in this case to a *historical decoupling between the first and second transitions or between two successive phases of legitimation processes*. These transitions, the first from the microlevel of action to the metastable or mesolevel of intermediary structures, and the second, from intermediary or metastable to macrostructures or reproducible institutions, connect a temporal and precarious innovation, which though real has no means to reproduce itself.

In this case study, historically authentic metastable states were identified for the three basic features of interaction, namely, communication of meaning, power relations and morality. Thus, the metastable state of meaning was identified with the semantic category of metaphor and specifically with the metaphor of "physico-chemical morphology." Metaphors fulfill the requirement of combining the contextuality of interaction with the objectivity of structures of signification because the discordant meanings of the metaphor's components (two or more words) maintain an unstable equilibrium with the unitarian meaning derivative of the symbolic action of its components upon each other (Black 1962, 1978; Geertz 1973; Kuhn 1979; Martin and Harré 1982; MacCormac 1985; Ortonyi [ed.] 1979; Ricoeur 1977; Sapir and Crocker [eds.] 1977; Turner 1971). The historically authentic metaphor "physico-chemical morphology" reflected throughout the 1930s both the contextuality of Needham and Waddington's interest in a new, theoretical and interdisciplinary approach to embryology and the potential objectivity of a new field. The metaphor's instability was revealed by the failure of policy to continue supporting the project: The metaphor practically ceased to exist.

With regard to morality, the metastable state was identified with the value of transdisciplinarity, a pluralistic and nonhierarchical form of scientific order, different from the disciplinary peck order and its tradition defined sharp boundaries. Transdisciplinarity persisted throughout the project's duration, sustained by an interventionist policy, but faded once the project was terminated. Transdisciplinarity remained a marginal phenomenon as long as the pre-World War Two era prevailed. In its aftermath, but especially in the 1960s, transdisciplinary collaboration became a highly valued form of scientific organization, once again, to a large extent as a result of various, new initiatives in policy, by then at the national and international levels (Abir-Am 1992a,b). It was not an accident that molecular biology finally obtained institutional legitimation precisely at the time when transdisciplinary forms of collaborations became not only legitimate but acquired hegemony over an increasingly mobile or changing scientific order.

Finally, with regard to power relations, the metastable state was identified with informal authority. The authority acquired by Needham and Waddington as a result of their new approach or mastery over the domain of physico-chemical morphology, was recognized for example by several professors at Cambridge and elsewhere, by the Rockefeller Foundation and its positive advisors, by the journals who published their papers and by the conferences and related professional fora in which they presented their work. But this was still informal authority because it was not tied to reproducible structures or positions, as was formal authority. The projected transition to formal authority was blocked once the project was terminated. In this sense, their informal authority grounded in the temporal structure of a transitory collaboration dissolved into contingent personal power once the collaborative project was terminated.

To conclude, this model of scientific change conceptualized failure as a (reversed) transition from objective or stable and semiobjective or metastable structures of signification, domination and legitimation to contingent and fluid action of individuals. On the one hand, the model emphasized the simultaneity of these transitions, both stabilizing and destabilizing, for the three basic properties of interaction, namely, communication of meaning, power and morality. This expands the explanatory range of the model beyond that of two-dimensional models revolving around social and cognitive features of science. Hence the model might be called the "me-so-po-tamian operon," so to evoke both the interdependence of meaning, social order and power and their mutually regulatory capacity, and its analogy to transdisciplinary models in molecular biology, such as the operon model of molecular regulation.

Second, the "me-so-po-tamian operon" emphasizes as a model the role of metastable states (of meaning, or metaphors; of power, or informal authority; of morality, or transdisciplinary collaboration) in instances of both successful and failed scientific change. Failure of scientific change is explained symmetrically to success, as a form of transition between different levels of stabilization of meaning, power and morality. Failure occurs when the destabilizing transitions take place, while success occurs when the stabilizing transitions complete the sequence from the microlevel of individual action to the macrolevel of social structure. Either way, the transitions occur simultaneously across the metastable level of intermediary systems of signification, or metaphors; intermediary systems of legitimation, or collaborative and transdisciplinary groups; and intermediary systems of power relations or informal authority.

The "me-so-po-tamian operon" has the most to offer for the analysis of historical events involving a prolonged decoupling of the transition from the contextual microlevel of scientific action to the transcontextual or objective macrolevel of scientific structures. It immediately offers a symmetrical interpretation of failure as a "reversed" transition or the transition from objective and semiobjective scientific structures to contextual scientific action. The model further enables the capturing of intermediate, metastable phenomena of signification, legitimation and authority formation, upon which any enduring change at the macrolevel is predicated. Having started with a problem of the role of policy failure in the early search for the "secret of life," (the code name of the discourse in molecular biology in the 1930s) and its subsequent historiography, we arrived at an incipient understanding of the "secret of social life" or the symmetry of social action leading to success and failure in historically situated scientific innovation.

Johns Hopkins University

Notes

*I am indebted to Larry Laudan for encouraging me to participate in the conference on "Scientific Failure" at the Center for Philosophy of Science at the University of Pittsburgh (April 1988) and to the conference organizers, Tamara Horowitz, Allan Janis and Gerald Massey for their generosity in allowing me to explore the dimensions of failure within my own favorite transdisciplinary combination of history of science, science policy, and social theory, in the company of bona fide philosophers of science. The analytical discussion of failure in this paper depended on prior archival documentation made possible by grants from the Wellcome Trust (London, UK), The Rockefeller Archive Center, and the National Science Foundation (DIR-89-22152). I also benefitted from comments by participants at several talks I gave on related topics, especially by Zev Bechler, John T. Edsall, and Camille Limoges.

References

Aaserud, F. (1990), *Redirecting Science, Niels Bohr, Philanthropy, and the Rise of Nuclear Physics*. New York: Cambridge University Press.

Abir-Am, P. G. (1980), "From Biochemistry to Molecular Biology: DNA and the Acculturated Journey of the Critic of Science Erwin Chargaff," *History and Philosophy of Life Sciences 2*: 3-60.

_____ . (1982a), "The Discourse on Physical Power and Biological Knowledge in the 1930s: A Reappraisal of the Rockefeller Foundation's Policy in Molecular Biology," *Social Studies of Science 12*: 341-382.

_____ . (1982b), "How Scientists View Their Heroes: Some Remarks on the Mechanisms of Myth Construction," *Journal of the History of Biology 15*: 281-315.

_____ . (1983/4), *The Biotheoretical Gathering in England, 1932-1938 and the Origins of Molecular Biology*. Ph.D. dissertation, Universite de Montreal.

_____ . (1984), "Beyond Apologetic History and Deterministic Sociology: Reassessing the Impact of Research Policy upon New Scientific Disciplines," *Social Studies of Science 14*: 252-263.

_____ . (1985a), "Recasting the Disciplinary Order in Science: A Deconstruction of Rhetoric on 'Biology and Physics' at Two International Congresses in 1931," *Humanity and Society 9*: 388-427.

_____ . (1985b), "Themes, Genres and Orders of Legitimation in the Consolidation of New Scientific Disciplines: Deconstructing the Historiography of Molecular Biology," *History of Science 23*: 73-117.

_____ . (1985-89; 90-91), "Research Schools of Molecular Biology in UK, US, and France, 1930-1970," proposals and reports to the National Science Foundation.

_____ . (1986), "A Policy of Unintended Consequences: The Role of Advisors, Officers and Grantees of the Rockefeller Foundation in the 1930s," talk given at the British Society for the History of Science summer meeting at Oxford (11 July 1986) and at the University of Colorado at Boulder (10 December 1986).

_____ . (1987), "The Biotheoretical Gathering, Transdisciplinary Authority and the Incipient Legitimation of Molecular Biology in the 1930s: New Perspective on the Historical Sociology of Science," *History of Science 25*: 1-70.

_____ . (1988a), "The Assessment of Interdisciplinary Research in the 1930s: The Rockefeller Foundation and Physico-chemical Morphology," *Minerva 26*: 153-76.

_____ . (1988b), "Recasting the Biological Order: Organismic Biologists Respond to the Rise of Molecular Biology," lecture delivered at Stanford University, 28 January 1988; forthcoming.

_____ . (1991a), "Nobelesse Oblige: Lives of Molecular Biologists," *ISIS 82*: 326-343.

_____ . (1991b), "The Philosophical Background of Joseph Needham's Work in Chemical Embryology," in S. Gilbert (ed.), *A Conceptual History of Modern Embryology*. New York: Plenum, pp. 159-80.

_____ . (1992a), "From Multidisciplinary Collaboration to Transnational Objectivity: International Space as Constitutive of Molecular Biology, 1930-1970," in E. Crawford, T. Shinn and S. Sorlin (eds.), *Denationalizing Science: The International Context of Scientific Practice*. Dordrecht: Kluwer Academic, pp. 153-186

_____ . (1992b), "The Politics of Macromolecules: Rhetoric, Biochemists, and Molecular Biologists," *Osiris 7*: 210-37.

Austoker, J. and Bryder, L. (eds.) (1989), *Historical Perspectives on the Role of the MRC*. Oxford: Oxford University Press.

Berger, P. and Luckman, T. (1967), *The Social Construction of Reality: A Treatise in the Sociology of Knowledge*. London: Sage.

Black, M. (1962), *Models and Metaphors*. Ithaca: Cornell University Press.

_____ . (1978), "More on Metaphor," *Dialectica 31*: 424-457.

Blume, S. S. and Spaapen, J. B. (1988), "External Assessment and 'Conditional Financing' of Research in Dutch Universities," *Minerva 26*: 10-30.

Branigan, A. (1981), *The Social Process of Scientific Discovery*. New York: Cambridge University Press.

Cairns, J.; Stent, G. S.; and Watson, J. D. (eds.) (1966), *Phage and the Origins of Molecular Biology*. Cold Spring Harbor: Cold Spring Harbor Laboratory Press.

Callon, M. (1986), "The Sociology of an Actor-network: The Case of the Electric Vehicle" in M. Callon, J. Law and A. Rip (eds.), *Mapping the Dynamics of Science and Technology*. London: Sage, pp. 19-34.

Chubin, D. E. (1987), "Designing Research Program Evaluation: A Science Studies Approach," *Science and Public Policy 10*: 82-91.

Chubin, D. E. and Jasanoff, S. S. (eds.) (1985), "Peer Review and Public Policy," *Science, Technology and Human Values 10*: 3-91.

Cohen, S. S., (1984), "The Biochemical Origins of Molecular Biology: Introduction," *Trends in Biochemical Sciences 9*: 334-336.

Collins, R. (1988), *Theoretical Sociology*. San Diego: University of California Press.

Cozzens, S. E. (1987), "Expert Review in Evaluating Programs," *Science and Public Policy 10*: 71-81.

Crick, F. (1988), *What Mad Pursuit: A Personal View of Scientific Discovery*. New York: Basic Books.

'Cueto, M. (1990), "The Rockefeller Foundation's Medical Policy and Scientific Research in Latin America: The Case of Physiology," *Social Studies of Science 20*: 229-54.

Dobzhansky, T. (1964), "Biology: Molecular and Organismic," *American Zoologist 4*: 443-452.

Edelman, G. M. (1988), *Topobiology: An Introduction to Molecular Embryology*. New York: Wiley.

Fournier, M.; Gingras, Y.; and Mathurin, C. (1988), "L'evaluation par les pairs et la definition legitime de la recherche," *Actes de la Recherche en Sciences Sociales 74*: 47-54.

Geertz, C. (1973), "Ideology as a Cultural System," in C. Geertz, *The Interpretation of Cultures*. New York: Basic Books, pp. 193-233.

Giddens, A. (1979), *Central Problems in Social Theory*. Berkeley and Los Angeles: University of California Press.

_____ . (1986), *The Constitution of Society: Outline of the Theory of Structuration*. Cambridge: Polity Press.

Gilbert, S. F. (ed.) (1991), *A Conceptual History of Modern Embryology.* New York: Plenum.

Hamburger, V. (30 May 1986) Letter to the author.

_____. (1988), *The Heritage of Experimental Embryology: Hans Spemann and the Organizer.* Oxford: Oxford University Press.

Horder, T. and others (eds.) (1986), *A History of Embryology.* Cambridge, England: Cambridge University Press.

Jasanoff, S. S. (1987), "Contested Boundaries in Policy-Relevant Science," *Social Studies of Science 17*: 195-230.

Judson, H. F. (1979), *The Eighth Day of Creation: The Makers of the Revolution in Biology.* New York: Basic Books.

Kay, L. E. (1986), "Cooperative Individualism and the Growth of Molecular Biology at Caltech, 1928-1953." Ph.D. Dissertation. Johns Hopkins University.

Kendrew, J. (1967), "How Molecular Biology Started," *Scientific American 216*: 141-143.

_____. (1970), "Remarks on the Origins of Molecular Biology," *Biochemical Society Symposia 30*: 5-10.

Kitcher, P. (1984), "1953 abd All That: A Tale of Two Sciences," *The Philosophical Review 93*: 335-73.

Knorr-Cetina, K. and Cicourel, A. (eds.) (1981), *Advances in Social Theory and Methodology: Toward an Integration of Micro- and Macro-Sociologies.* London: Sage.

Kohler, R. E. (1991), *Partners in Science: Foundations and Natural Scientists, 1900-1945.* Chicago: Chicago University Press.

Kuhn, T. S. (1959), "Metaphor in Science," in A. Ortony (ed.), *Metaphor and Thought.* Cambridge, MA: MIT Press, pp. 409-19.

Lwoff, A. Ullmann, A. (eds.) (1979), *Origins of Molecular Biology: A Tribute to Jacques Monod.* New York: Academic Press.

Lynch, M. (1985), *Art and Artifact in Science: Work Talk and Work Shop in a Neuroscience Laboratory.* London: Routledge.

MacCormac, E. R. (1985), *A Cognitive Theory of Metaphors.* Cambridge, MA: MIT Press.

Martin, J. and Harré, R. (1982), "Metaphor in Science" in D. S. Miall (ed.), *Metaphor: Problems and Perspectives.* London: Sage, pp. 89-105.

McCarty, M. (1985), *The Transforming Principle Discovering that Genes are Made of DNA.* New York: Norton.

Monod, J. (1969), "From Molecular Biology to the Ethics of Knowledge," *The Human Context 1*: 325-336.

Monod, J. and Borek, E. (eds.) (1971), *Microbes and Life.* Ithaca: Cornell University Press.

Needham, J. (1946), *History is on Our Side: A Contribution to Political Religion and Scientific Faith.* London: Heinemann.

_____. (1968), "Organizer Phenomena after Four Decades: A Retrospect and

Prospect" in K. Dronamraju (ed.), *Haldane and Modern Biology*. Baltimore: The Johns Hopkins University Press, pp. 277-97.

Needham, J.; Waddington, C. H.; and Needham, D. M. (1934), "Physico-chemical Experiments on the Amphibian Organizer," *Proceedings of the Royal Society B, 114*: 393-422.

Offer, A. (1989), *The First World War: An Agrarian Interpretation*. Oxford: Oxford University Press.

Olby, R. (1974), *The Path to the Double Helix*. London: Macmillan.

Perutz, M. (1980), "Origins of Molecular Biology," *New Scientist* (January 31) *11*: 326-329.

Porter, A. L. and Rossini, F. A. (1985), "Peer Review of Interdisciplinary Research Proposals," *Science, Technology and Human Values 10*: 33-38.

Prigogine, I. and Stengers, I. (1984), *Order out of Chaos: Man's Dialogue with Nature*. London: Heinemann.

Rich, A. and Davidson, N. (eds.) (1968), *Structural Chemistry and Molecular Biology*. San Francisco: Freeman.

Ricoeur, P. (1977), *The Rule of Metaphor*. Toronto: University of Toronto Press.

Sapir, J. and Crocker, J. (eds.) (1977), *The Social Use of Metaphor: Essays on the Anthropology of Rhetoric*. Philadelphia: University of Pennsylvania Press.

Schutz, A. (1962), *The Problem of Social Reality*. The Hague: Nijhoff.

_____ . (1965), *Studies in Social Theory*. The Hague: Nijhoff.

_____ . (1967), *The Phenomenology of the Social World*. Evanston: Northwestern University Press.

Stent, G. S. (1972), "Prematurity and Uniqueness in Scientific Discovery," *Scientific American 227*: 84-93.

_____ . (ed.) (1980), *The Double Helix and Its Reviews*. New York: Norton.

Turner, V. (1964), "Betwixt and Between: The Liminal Period in Rites de Passages," *Proceedings of the American Ethnological Society 16*: 4-20.

_____ . (1971), *Dramas, Fields and Metaphors: Symbolic Action in Human Society*. Ithaca: Cornell University Press.

Waddington, C. H. (1963), "Two Cheers for the Organizer," *Nature 198*: 42.

_____ . (1969), "Some European Contributions to the Prehistory of Molecular Biology," *Nature 220*: 318-322.

_____ . (1975), *The Evolution of an Evolutionist*. Ithaca: Cornell University Press.

Watson, J. D. (1968), *The Double Helix*. New York: New American Library.

Werskey, G. (1978), *The Visible College*. London: Lane.

Wyatt, V. (1972), "When does Information become Knowledge?" *Nature 235*: 86-89.

Zallen, D. T. (1992), "The Rockefeller Foundation and Spectroscopy Research: The Programs at Chicago and Utrecht," *Journal of the History of Biology 25*: 67-89.

8. THE DETERMINANTS OF A SCIENTIST'S CHOICE OF RESEARCH PROJECTS*

Arthur M. Diamond, Jr.

Scientists may differ in their choice of research projects due to differences in tastes (e.g., degree of risk aversion) and due to differences in constraints (e.g., ability, time, laboratory resources, information on likelihood of success, etc.). Although Stigler in his Nobel Lecture (1983) has suggested that conservative scientific behavior is due to risk aversion and to a desire to protect human capital already invested in current theory, economic explanations of the behavior of scientists are at a fairly early stage of development. The aim of the current research is partly to develop the economic theory of scientific behavior, but mainly to learn what the stylized facts are upon which future theory should focus.

For the present study, limited data have been collected on approximately 200 scientists in the West who either wrote on polywater or else were members of the population of scientists who could have written on polywater. "Polywater," condensed in very small capillaries, was thought to be a new form of water that had a lower freezing temperature and a higher boiling temperature than ordinary water. In the West the surge of articles began in 1969. By 1973 when everyone admitted that the polywater phenomena were due to impurities in normal water, over 100 academic papers had been written on the subject.

More detailed biographical and professional data have been obtained on a subsample of about 100 scientists from the larger sample. In addition, a survey was sent to all of those in the larger sample for whom any potential mailing address was available. Using the evidence from the polywater episode, I examine several hypotheses concerning alleged determinants of a scientist's choice of research projects. In particular, I briefly examine risk aversion, age, number of children, income and academic tenure, funding and journal demand, religion, geographical location, characteristics of family of origin, and number of citations.

1. THE PROBLEM

Since Kuhn's *The Structure of Scientific Revolutions* (1970), considerable attention in philosophy and sociology of science has focused on how much of scientists' decisions on theories can be explained in terms of the objective criteria emphasized in textbooks ("internal" factors) and how much needs to be attributed to prejudice, ideology, age and economic interest ("external" factors).[1] Accepting the common, but disputed interpretation of Kuhn's book that external explanations of decision making undermine science's claim to objectivity and progress,[2] many philosophers (e.g., Toulmin 1972; Laudan 1977; Hull 1978; Diamond 1978, 1988b), have attempted to reconcile the internal and external accounts in a way that does justice to the history of science while still defending something of science's claim to special epistemic status.

While economists have attempted to model the labor market for scientists, historians of science and scientists themselves have continued to author accounts of scientific decision making that place varying emphasis on internal *versus* external factors.[3] For example, since Watson's (1968) now classic account, some scholars have discussed the DNA episode with an emphasis on external explanations (see, e.g., Toulmin 1972, Fleming 1968), while others have emphasized internal explanations (see, e.g., McCarty 1980, Diamond 1982). Case studies of this sort are often useful and sometimes provide the best evidence that one can reasonably hope to obtain about the reasons for the acceptance or rejection of historically important scientific theories. But sometimes it is possible to do better. In particular, certain sorts of external factors allegedly important for theory acceptance are quantifiable. In such cases rigorous social science statistical techniques allow us to determine how much of the differences in acceptance of a theory are explainable in terms of differences in the quantifiable factors. The larger the residual variance in acceptance after the quantifiable external factors have been taken into account, the larger the potential role for purely internal factors to play. More sophisticated inferences are also possible. Failure of the quantifiable external factors to be important might lead one to look with renewed interest at the nonquantifiable external factors. Alternatively, even if quantifiable external factors turn out to be statistically important, it can be argued that through institutional natural selection or some other process, at least some "external" quantifiable factors are correlated with objective "internal" theory selection procedures (see, e.g., Hull 1978, Diamond 1978, 1988b). An example might be high salaries (an external factor) being correlated with skill at judging a new theory's explanatory power (an internal factor). Although not eliminating all disputes in metascience, rigorous testing of the effects of quantifiable external factors thus has the positive effect of improving our knowledge of the phenomena that any plausible metascientific position must explain.

Work in the Merton tradition of sociology of science has in the last twenty years vastly increased our systematic knowledge of the workings of scientific institutions (Cole and Cole 1973, Zuckerman 1977, Crane 1972, Goffman and Warren 1980, Hargens 1988; Mullins *et al.* 1977, Studer 1980). But although much has been learned about the determinants and correlates of success in paper-writing, citations, and professional advancement, little has yet been done to systematically increase our knowledge of the determinants and correlates of theory acceptance. Instead, discussions of theory acceptance usually have taken the form either of abstract generalization or else of detailed narrative case studies. Zuckerman's article (1979) provides a good survey of much of this work.

2. AN ECONOMIC APPROACH TO THE PROBLEM

Assume that the objective of a scientist is to advance the frontier of knowledge either because this objective is a direct source of utility for the scientist or because it is a productive activity for which the scientist is compensated. Assume further that the choice variable for a scientist is the research project. Research projects may differ along several dimensions, most notably the probability of success, the time to completion and the magnitude of the contribution if successful. Broadly speaking, differences in choice of research projects will depend either on differences in values or on differences in constraints (or on some combination of the two). Under the heading of differences in values, one frequently hears that scientists differ in the extent to which they are risk averse.

Under the heading of constraints, the following would presumably be important:

1. The scientist's belief about whether a project can be successfully completed.

2. The scientist's belief about whether *she* can successfully complete the project.

3. The scientist's belief about how important the contribution to science will be if the project is successfully completed.

4. The scientist's beliefs about all of the above for other research projects.

The above beliefs would themselves depend on the objective state of scientific knowledge, the information that the scientist has on the state of scientific knowledge, the scientist's ability, the scientist's resources and the scientist's beliefs about her ability and resources. Although economists are apt to deny it, other social scientists frequently have concluded that persons with the same ability and the same resources may nonetheless differ in their estimate of the probability of their own success. Such differences would be attributed to differences in "self-confidence." My discussion assumes the subjectivist interpretation of probability. The assumption is increasingly supported by current work on the frontiers of statistics indicating that "randomness" is not a feature of the objective world (Kolata 1986).

Although considerable exploratory research has been done on the behavior of scientists, the subject is still in the early stages of development. Much additional research will be required before we can say with confidence which constraints matter most in determining the choice of research projects and whether there is a residual of choice behavior that is due to differences in risk aversion.

The objective of the current research is mainly to provide additional information on the characteristics of those who chose and those who did not choose a particular well-defined research project. The hypotheses to be examined are suggested by many sources both within economics and outside of economics. But all the hypotheses can be viewed as fitting into the framework just sketched, that is, they are related to the scientist's ability, to the scientist's other resources, to the scientist's knowledge of herself, to the scientist's knowledge of science and to the scientist's taste for risk.

The well-defined research project that I will examine is the polywater episode. I first briefly summarize the episode and describe the data set. Then I examine some hypotheses about what characteristics we would expect to be correlated with the choice of an ex ante promising, but ex post mistaken, research project. I will locate the hypotheses within the choice model sketched above. Finally, I will examine how the evidence from the polywater episode bears on the truth of the hypotheses.

3. THE POLYWATER EPISODE

Polywater, also known as 'anomalous water,' was first discovered in 1961 by an obscure Russian scientist named Fedyakin who found that in very small capillaries a form of water could be condensed that has a lower freezing temperature and a higher boiling temperature than ordinary water. The discovery was taken up by the prestigious Russian scientist Deryagin who brought his research to the attention of English-speaking scientists in a series of lectures from 1966 to 1968. In the West the surge of articles on polywater began in 1969. By 1973 when Deryagin admitted that the polywater phenomena were probably due to impurities in normal water, over 400 publications had appeared that referred to the subject (see Bennion and Neuton 1976).

Among mistaken research projects, some are only mistaken *ex post*, while others are mistaken both *ex post* and *ex ante*. Classifying a particular project is often difficult, but the polywater case to be studied here, as well as the N-Ray episode (Klotz 1980), phrenology (Cantor 1975) and some of the cases discussed by Langmuir (1989) are probably best thought of as *ex post* mistakes. Good examples of research projects that were mistaken both ex post and ex ante can be found in some of the chapters of Wallis's *On the Margins of Science* (1979).

Franks's *Polywater* (1981) is by far the most extensive account of the episode but other useful accounts are also available (Hasted 1971, Howell 1971, Gingold 1973, Gingold 1974). The polywater episode has several advantages over other mistakes that could be studied: (a) a consensus exists that, at least *ex post*, research on polywater was a mistake, (b) the episode occurred recently enough for data to be available in the *Science Citation Index* on citations and number of articles, (c) the episode occurred long enough ago to permit observation of career consequences for the polywater scientists and (d) the number of identifiable scientists involved was larger than for most other mistakes that could be studied.

Kollman and Allen (1972) estimate that by January 1972 about 25 theoretical articles had been written on polywater. The development of theoretical models of anomalous water before the existence of the phenomena was well established is one of the aspects of the episode that has been criticized as being in violation of accepted standards of scientific practice. Allen and Kollman, to the contrary, in a theoretical article coauthored with others, aggressively defended early theorizing by saying that:

> Many scientists question the existence of such a new and well-defined form of water and it is therefore of obvious importance to investigate the problem theoretically— preferably by *ab initio* techniques. (Sabin *et al.*, 1970, 235)

What was obvious to Allen and Kollman in 1970 was apparently no longer so obvious in 1972 after the scientific community was close to a consensus that the polywater phenomena were due to impurities in normal water:

> It is not unexpected that a number of those theoreticians active in the theory of ordinary hydrogen bonds participated in the polywater problem, but the manner of their participation has been unusual—it is only rarely that both theoreticians and experimentalists have simultaneously been involved during that phase of research when the existence as well as the properties of a new material have been in question. (Kollman and Allen 1972, 288)

Franks has gone so far as to suggest that one methodological lesson of the episode is, quoting Sherlock Holmes, that "it is a capital mistake to theorize before one has data" (1981, 87). (In the early stages of the controversy, however, Franks had been more sympathetic to the pro-polywater researchers [1968, 48-49]).

Some current commentators are ready to label the polywater episode as bad science (see Eisenberg 1981; and Franks 1981, 180-182). Zuckerman, for instance, discusses the polywater episode as a "classic" case of a "disreputable error" because polywater's proponents failed "...to live up to the cognitive norms prescribing technical procedures designed to rule out even the most favored hypotheses if they are in fact unsound" (1977, 111 and 112). But other scientists (e.g., Deryagin 1983; Gould 1981, 15; and Pethica 1982) seem to view polywater more generously as one of those "...mistakes of good men earnestly seeking the truth" (Metzger 1970, 9).

Certainly there must be a continuum of mistakes graded by severity with plagiarism and data falsification on one extreme and contextually plausible, but ultimately unfruitful, hypothesis on the other. In another place, Diamond (1986a), I discuss in considerable detail the question of whether the pro-polywater theoreticians and experimentalists were doing work that was consistent with the then-accepted standards for doing good science. My conclusion in that paper is that the better of the experimentalists and probably the better of the theoreticians were indeed doing work that met the then-accepted standards. If I am correct, the implication is that polywater was an ex post, but not an ex ante mistake.

However, for the purposes of this paper I need not precisely locate pro-polywater research on the continuum of mistakes. No matter how severe a mistake polywater is ultimately judged to have been, it is still important to understand why some scientists chose polywater as a research topic.

4. THE DATA

The data set compiled for this study is 'longitudinal,' which means that information is obtained on individuals over an extended period of time. The advantages of longitudinal data in studying scientists are discussed by Long (1978) and Diamond (1986b).

Biographical and career information was obtained for a sample of scientists who wrote on polywater (call them the 'polywater scientists') and a comparable sample of scientists who did not write on polywater (call them the 'non-polywater scientists'). Gingold's (1973) nearly exhaustive bibliography on polywater was the basic source for my sample of polywater scientists. Gingold's aim was apparently to include all publications, regardless of language, that mention polywater in any way. Most of the bibliography consists of publications that had appeared before 30 October 1972, although a few entries were added on the final page while Gingold's article was in proof.

As a partial check of the exhaustiveness of the Gingold bibliography, I compiled a list of entries that appeared in the *Science Citation Index (S.C.I.)* either under the heading 'anomalous water' or under the heading 'polywater' for the years 1968 through 1972, inclusive. I found that of the 97 entries, 81 were on the Gingold list. Although most of the remaining entries have both 'water' and 'anomalous' in the title, they do not appear in any sense to be related to the polywater research. I interpret this as evidence that Gingold's bibliography is a reasonably complete listing of the important research articles on polywater from 1968-1972.

I also eliminated monograph and textbook entries from my sample on the grounds that in modern times current research in science is reported in articles (see Kuhn 1970, 19-20). A few articles that appeared in edited books were included. Of the 437 total entries on the bibliography, I examined copies of 325.[4] Books, working paper technical reports and articles in foreign language journals account for most of the entries for which photocopies were not obtained. Of the 325 entries obtained, I excluded from further analysis those publications that were judged not to be research articles on polywater. The excluded publications consisted mainly of either letters, review essays, articles for the popular press, or articles written in some language besides English.

Some journals' letters sections contained contributions that represented more substantive research than other journals' notes sections. So, although in general, notes were included in the sample and letters were excluded, a few exceptions were made. I also excluded research articles on subjects other than polywater that mention polywater only as an aside without bringing forward new evidence or argument. More specifically, if an article assumes the existence of polywater in order to make a substantive argument on another issue (e.g., Gingold # 171), then it was included in the sample (and classified as "pro"). If it briefly and vaguely mentions polywater as possibly relevant to the issue under discussion (e.g., Gingold # 421), then the article was excluded from further analysis. After the elimination of all excluded entries, my final sample of research publications on polywater consisted of 112 entries.

These remaining entries were examined in order to judge whether their attitude toward polywater was pro, con or neutral. An article was judged as being pro-polywater if on balance it seemed to support the position that the anomalous phenomena were due to some different structure of water. An article was judged as being con-polywater if on balance it seemed to support the position that the anomalous phenomena were due to impurities in the water. Articles that were so heavily qualified that they did not seem to lean toward either position were judged as being neutral. The process of judging an article's position is made more difficult by standards of scientific style that encourage qualification and discourage speculation or direct criticism of the work of others (see Diamond 1982, and Mac-Roberts and MacRoberts 1984).

A scientist was judged as having been pro-polywater if she had ever authored or coauthored a research article that was judged to have been pro-polywater. Thus Allen and Kollman who wrote an important pro article in 1970 and an important con article in 1971 were classified as 'pro.' A scientist was judged as having been con-polywater if she had never written a pro research article but had authored a con research article. A scientist was judged to have been neutral on polywater if she had never authored either a pro or con research article, but had authored an article that was judged to have been neutral.

The 437 entries in the Gingold bibliography were written by 415 different authors. The numbers differ because many entries had more than one author and some authors wrote more than one entry. Of the 415 authors, I was able to identify 115 authors whose only contribution to the bibliography consisted of publications that were excluded from the article sample for any of the reasons mentioned above. The 115 were excluded from the sample analyzed later in the paper.[5]

I also excluded from the data analyses 61 non-Western scientists (mainly Soviets) on the grounds that the citation process and reward structure in the East may have differed from those in the West. A second reason for excluding Soviets from the sample is that the plethora of Western transliterations of Soviet names substantially increases the noise in citation counts for Soviet scientists. I excluded as Soviets those (1) who were identified as Soviets by Franks or (2) who wrote only in Soviet sources *and* were not identified in any of my biographical sources as having been employed in the West from 1968-1983. Thus George K. Swinzow who coauthored with Deryagin in a Soviet journal and who grew up in the Soviet Union is *not* counted as a Soviet by us because he was employed by United States laboratories throughout the period.

Finally, another 44 authors were excluded because I had not been able to obtain any of the publications by them. A couple of scientists, for example, S. Levine and Joe L. White, were excluded because their last names and first initials were the same as those of other scientists in related fields, implying that distinguishing their citations from the citations of the like-named scientists would be prohibitively difficult (the *S.C.I.* does not give full first names in the citation volumes).

Using the article evaluations, I was able to code the remaining 195 scientists as being either pro, con or neutral. Of the 112 scientists classified as pro, 101 wrote at least one article that was pro and none that were con while 11 wrote at least one article that was pro and at least one article that was con. The 78 scientists classified as con each wrote at least one article that was con and none that were pro. The 5 scientists classified as neutral each had written only articles that were neutral. The classification of the 415 authors on the Gingold list into pro, con, neutral and various excluded categories is summarized in Table A-1. All tables in the paper are grouped together at the end of the text, just before the footnotes.

A control group was selected of scientists who presumably could have chosen to write on polywater but did not. The identification of those in the control group is difficult in part because the polywater topic stimulated considerable interdisciplinary interest. Bennion and Neuton have suggested that:

> Because of its startling implications, the possibility of a polymeric form of water caught the fancy of scientists worldwide in many disciplines, especially biology, chemistry, physics and the bio-medical areas. A conservative estimate would place the number at no less than 10^5 for those having the immediate potential of becoming involved in polywater research. (Bennion and Neuton 1976 54)

The distribution of polywater scientists by discipline is presented in Table A-2. Note that of those identifiable by discipline, over 60% had some affiliation with chemistry. The next most important discipline, physics, had only roughly 10% exclusively affiliated with it. I decided to restrict the control group to chemists in part because chemists had been the dominant group in polywater research and in part because chemists document themselves (through their *Directory of Graduate Research*) better than do physicists. I restricted the control group to those who had published in *Science* because that was the most important vehicle for Western polywater research and because I wanted to insure that the control group consisted (like the polywater group) of those who were active in publishing. More concretely, the control group consists of the 37 authors of Research Notes (also sometimes called "Research Reports") in *Science* from 2 January 1970 through 19 March 1971 who, from their listed affiliations, appeared to be chemists. Some evidence that the control group does indeed contain those who could have written on polywater but chose not to, is provided by the three scientists who would have qualified for the control group but were excluded because they were also polywater scientists. The three excluded scientists were Kamb, Pimentel and Turkevich.

For all scientists in the sample, first-authored citation counts were obtained from the *S.C.I.* for the years 1968, 1969, 1974 and 1981. The *S.C.I.* only lists citations under the first author of an article. As a result, full citation counts can only be obtained by learning the first authors' names of all coauthored papers that a scientist has written and then obtaining a citation count for each paper from the *S.C.I.* under the first author's name. The process is time-consuming and laborious, the more so where, as in chemistry, the average number of articles is large and coauthorship is common. Some evidence (Diamond 1986b) suggests that first author citation counts may be a reasonably good proxy for full citation counts. For the present only first author counts have been used.

Values for biographical variables were obtained either from the various bi-annual editions of the *Directory of Graduate Research* (1977), from the first edition (1984) of *Chemical Research Faculties: An International Directory*, from the various editions of *American Men and Women of Science* or from responses to the survey. Data on number of publications were obtained from the annual source volumes of the *S.C.I.* Data on number of citations per year to the scientist's work were obtained from the annual citation volumes of the *S.C.I.*

Descriptive statistics for the various subsamples appear in Tables A-3 through A-5. Table A-6 identifies the scientists who contributed 6 or more publications to the polywater literature. Table A-7 provides variable definitions and Table A-8 presents simple correlations for many of the key variables in the study. In each cell in Table A-8, the first number is the simple correlation coefficient; the second number is the significance probability for use in testing the hypothesis that the correlation coefficient has a statistically significant difference from zero; and the third number is the number of observations used in calculating the correlation coefficient. The correlation coefficient is conventionally judged as "statistically significant" if the significance probability is less than 0.05.

Henry Small of the Institute for Scientific Information has graciously provided me with citation cluster data that were calculated by the Institute (see Garfield 1983, Mullins *et al.* 1977, and Small and Griffith 1974). The data identify those polywater scientists who were most visible in their acceptance of polywater and, hence, whose careers might have been most damaged. The articles that are included in the pro and con clusters are listed in Tables A-9 and A-11, respectively. The names and some descriptive statistics for the authors of the articles in the pro and con clusters are given in Tables A-10 and A-12, respectively.

A survey was conducted of the scientists in this study. The initial mailing was sent to the most recent known address of all those authors living in the United States who either had written publications listed in the Gingold bibliography or else were members of the control group. In the course of the paper, I will quote remarks made by the scientists in their answers to the survey. Since the scientists were assured confidentiality, I will refer to them only by their general position (i.e., pro, con, etc.). The survey response rates for all of the subsamples are given in Table A-13.

Most of the survey questions that will be utilized here were of the sort that asked the respondent for factual information about past or present demographic, family background and income variables. As such, they were similar in kind to questions asked in surveys routinely used by labor economists such as the National Longitudinal Surveys. A couple of the questions, however, were of the sort that economists are generally skeptical toward. In particular, I asked the scientists if they had viewed polywater research as risky when they made their decision to choose it as a research topic. McCloskey (1985, 9-10) discusses the usual story that has been told to discredit such questions, but goes on to note that the story might be given a new moral: Some questions about motives and beliefs are more useful than others. Even Friedman, who is one of those most opposed to the use of such questions for testing, argues that they often are useful as a source of hypotheses (1953, 31).

In my survey, I asked the polywater scientists whether their papers had supported the existence of an anomalous form of water. As mentioned earlier, I also examined the papers independently in order to judge from the published evidence whether the scientists had supported the existence of an anomalous form of water. Three of those whom I had judged to have been pro, considered themselves con. None of those whom I had judged con considered themselves pro. Also worth noting is that several of the others who were pro made a point of minimizing the extent of their involvement (in time and effort) in polywater research.

5. THE HYPOTHESES

I discuss several broad classes of hypotheses that are relevant to the account of research project choice that was sketched above. The discussion will include a description of the hypothesis and remarks on how it relates to the general account of research project choice. The discussion of each hypothesis will conclude with evidence from the polywater episode on the truth of the hypothesis. The hypotheses are divided into two broad groups. First, I consider those that are mainly related to how scientists differ in the extent of their risk aversion. This group includes attitudes toward risk, age, number of children, income and academic tenure. Second, I consider those that are mainly related to their information, ability and other constraints. This group includes funding and journal demand, religion, geographical location, and characteristics of family of origin and number of citations (as a measure of knowledge at the frontiers of science). Some of the hypotheses could be included in either group, depending on which version of the hypothesis is emphasized. Age, for instance, may affect risk aversion, but may also affect information and ability.

6. ATTITUDES TOWARD RISK

Stigler in his Nobel Lecture (1983, 538) has suggested that different attitudes toward risk can partially explain why some scientists accept a new theory while others reject it. The few scientific innovators are not risk averse, according to Stigler. The vast majority of scientists, including most of the young, are risk averse, and hence tend toward scientific conservatism.

Lester Telser has argued that variation in product quality will be less for well-advertised products than for other products, since consumers will be more apt to recall and retell bad experiences with a well-advertised product (Telser 1974, 32). Similarly, one might argue that variation in article quality would be less for a well-cited scientist than for others since scientists are more likely to recall and retell having read a bad article by a well-cited scientist. The well-cited scientist thus acts more risk averse than the less-cited because the well-cited scientist has more to lose from a mistake and less to gain from a success.

Psychologist David C. McClelland has argued that persons with a high need to achieve (as measured by his battery of test questions) have greater general self-confidence (1976, 222-223). Thus in situations in which they have limited prior information they will estimate that their chances of success are greater than will otherwise similar persons with a low need to achieve. When good information on past performance on similar tasks is available, however, both those with a high need to achieve and those with a low need to achieve are equally accurate in estimating their chances of success. Thus under conditions of uncertainty about personal performance, persons with a high need to succeed will take greater risks than persons with a low need to succeed.

The pro polywater scientists seem to have differed in how they perceived the riskiness of the polywater episode. I asked the scientists whether they considered polywater a very risky topic, greater than average riskiness, about average in riskiness or safe. The responses to this question, broken down by subsample, are presented in Table A-14. One notable feature of the answers is that a majority of the pro scientists report having perceived the polywater project to have been above average in riskiness.

In retrospect, several more precise questions might have been more informative than the broad one actually asked. For example, one question might have been how high the scientists thought that the probability was that the research might not be fruitful. A separate question might have been whether the scientists thought that, if unsuccessful, the research would damage their careers. Several of the scientists in the sample worked on the project because they were post-doctoral fellows or research assistants for a more distinguished colleague. One of these in fact told Franks (1981, as quoted on p. 190) that "...as a graduate student I don't stand to lose as much if I turn out completely wrong."

An additional ambiguity in the original question applies solely to the scientists in the con and *Science* subsamples. For them the question could have been interpreted as asking about the risk of writing a pro article on polywater or, alternatively, of writing on polywater at all (pro or con). Writing con on polywater in 1973, for instance, would have been a very low risk activity.

The ambiguity of the question on risk for the con and *Science* subsamples is reduced, however, by considering the responses of the *Science* subsample to the question: If you were aware of polywater research at the time, why did you not choose polywater as a research topic? All of the responses for the *Science* sample are reproduced below:

(500) I had better things to do.

(506) Too high a probability it wasn't a real phenomenon.

(511) Other, more interesting things to do.

(516) Not close to my field.

(526) Couldn't think of any good experiments or novel approaches.

(529) 1. The reports as presented were not credible.

2. In my opinion, the definitive experiments required very sophisticated equipment, not available to me at that time.

3. My research group was fully occupied with more significant problems.

(534) Good luck.

7. AGE

According to Edge, sociologists expect that greater risks will be taken by younger scientists than older scientists (Edge 1977, 336).

The gene-maximization hypothesis from sociobiology has been used in a couple of papers by economists in order to make inferences about optimal risk aversion in various circumstances (Rubin and Paul 1979, Diamond and Locay 1989). Rubin and Paul, for instance, argued that societies would have a survival advantage in which (holding wealth constant) mature males had greater risk aversion than adolescents.

Max Planck has claimed that: "...a new scientific truth does not triumph by convincing its opponents and making them see the light, but rather because its opponents eventually die, and a new generation grows up that is familiar with it" (Planck 1949, 33-34). Nearly all those who quote Planck's Principle accept it as true. The usual inference drawn from the Principle is that science is an irrational enterprise, "rationality" here being understood as the adherence to objective standards in judging theories.

Although the inference is *usually* drawn, I have shown elsewhere (Diamond 1988) that, even if older scientists do in fact accept new theories more slowly, this phenomenon is compatible with all scientists evaluating theories on the basis of standards such as elegance, rigor, and explanatory scope. Whether or not such an argument succeeds in making Planck's Principle compatible with the rationality of science, the first question is: Was Planck right? Until several years ago, the truth of his principle was merely assumed, but recently a few studies have begun to test it.

Finding important scientists who are counterexamples to Planck's Principle is one way to test the truth of the principle (Brush 1976, 94; Hagstrom 1965, 291; and Blackmore 1978, 347-349). But such counterexamples can never serve as conclusive refutation of the principle since they leave open the possibility that the Principle still holds for scientists most of the time. More rigorous and systematic tests of the importance of a scientist's age as a determinant of theory acceptance can be found in Hull *et al.* 1978, Diamond 1980, and Gieryn and Hirsh 1983. All three studies found a statistically significant effect of age in the direction predicted by Planck, although the first two found that the magnitude of the effect was small.

We are interested here in learning the effect of age on the acceptance of polywater. I have data on year of Ph.D. for many more scientists than I have data for year of birth. So instead of looking at the effect of age on the acceptance of polywater, I look at the effect of experience (which is highly correlated to age and could be alternatively labelled as "professional age"). The operational definition of "experience" is 1971 minus the year of receipt of the Ph.D degree. Even apart from data considerations, experience may be the more appropriate concept if I accept the 'costs of retooling' explanation for Planck's Principle.

In results reported in Diamond 1988a, I find that experience was not a statistically significant determinant of the probability that a scientist would accept polywater.

8. Number of Children

The number of children that a scientist has may effect her scientific decisions in a number of ways. One is that having dependent children may make a scientist more risk-averse (see Rubin and Paul). Another is that children and advancing the frontier of knowledge may be substitute technologies for producing the 'impact on future generations' z-good.

Evidence on the number of children reported by respondents to the survey is reported in Table A-16. No systematic differences between the different subsamples in average number of children are suggested by the table.

9. Income

Two measures of risk aversion have become standard in the economics profession: absolute-risk aversion and relative-risk aversion. Where utility is a twice differentiable function of wealth with positive first derivative and negative second derivative, absolute-risk aversion is defined by:

$$8.1 \quad \frac{U''(W)}{U'(W)}$$

and relative-risk aversion is defined by:

$$8.2 \quad \frac{WU''(W)}{U'(W)}$$

where U is utility and W is wealth. Economists routinely assume that absolute-risk aversion is a decreasing function of wealth. A less routine assumption is that relative-risk aversion is an increasing function of wealth (Arrow 1971, 96).

Apart from the just-mentioned hypotheses, economists have seldom ventured to hypothesize how persons can be expected to differ in their extent of risk aversion. Holding wealth constant, differences in risk aversion between persons are usually attributed to differences in tastes. As with other tastes, economists generally adopt an agnostic position, assuming (again holding wealth constant) either that risk aversion is constant among persons or else that it varies stochastically.

One notable exception to this generalization is a 1983 paper by Morin and Suarez. Using Canadian data on household assets, they find that relative-risk aversion is a *decreasing* function of wealth. In addition, they find that when wealth is controlled for, relative-risk aversion increases with age. A second exception to the economist's usual agnosticism on risk aversion is Reuven Brenner who concludes that "...when one's relative position in the distribution of wealth is worsened, one will gamble more frequently on new ideas..." (1983, 27).

Income in 1971 for scientists in each of the subsamples is reported in Table A-16. The accuracy of the data may be limited to some extent by the fact that the scientists are being asked to remember their income over 10 years earlier. No systematic differences in income are apparent for the different groups.

10. TENURE

A few labor economists in recent years have attempted to model the labor market experience of academic scientists (Freeman 1977, Siow 1979, Harris and Weiss 1981, Weiss and Lillard 1982).

For Siow, productivity is exogenous although, unlike Freeman, all scientists draw from the same productivity distribution. All scientists in the first period receive less than their expected value of marginal product and all scientists in the second period receive more than their expected value of marginal product. Siow's formal model has nothing to say about a scientist's choice of research topics although he writes some suggestive obiter dicta near the end of the paper. In particular, Siow hypothesizes that younger pre-tenure scientists will take fewer risks than older post-tenure scientists (Siow 1979, 20). Remarks by Weiss and Lillard (1982) seem to support the Siow hypothesis.

More recently, other papers which have attempted to explain the existence of tenure have been written by Ito and Kahn (1986), Carmichael (1988), and Kahn and Huberman (1988).

I have no data on the year of receipt of tenure for any of the members of my sample. However, I do have longitudinal data on academic rank for many scientists in the sample. Since year of promotion to the rank of associate professor usually corresponds closely to the year of tenure, such information should be useful in evaluating the effect of tenure on a scientist's position on polywater. Table A-17 reports the rank of scientists in various subsamples as of the beginning of the polywater episode in the West (i.e., 1969). Most of the scientists in the sample had obtained the rank of associate or full professor by the beginning of the polywater episode. It appears that slightly more scientists in the con and *Science* subsamples did not have tenure than in the pro subsample. This would be consistent with Siow's hypothesis that after the achievement of tenure, scientists choose riskier research projects.

11. FUNDING AND JOURNAL DEMAND

We would expect that, ceteris paribus, scientists would choose research projects that could be funded and would result in publishable articles. Several of the survey respondents suggested that funding and an interest by prestigious journals in the polywater episode explains why many scientists chose polywater as a research topic. One of the pro scientists (41) said in the questionnaire that:

> Most people writing about polywater or who sought research grants did so simply because the funds were available, not because they may or may not have believed in the existence of polywater.

12. RELIGION

Merton ([1936] 1970) in his doctoral dissertation, adapting earlier work by Weber ([1930] 1958), suggested that the Protestant ethic encouraged good science. Hardy (1974) has suggested that certain denominations seem to be over-represented among scientists relative to their numbers in the general population. The interpretation of simple correlations between religious denomination and measures of scientific achievement must be undertaken with great care because denomination may be a proxy for a wide range of other variables including, for example, income and family size.

Information on the religious affiliation of the scientists in the sample is presented in Table A-18. No systematic differences in religious affiliation appear to be present across the subsamples.

13. GEOGRAPHICAL LOCATION

Geographical location may be important to a scientist's choice of research projects for several reasons. A scientist at a major research center will likely have access to important resources required for successfully carrying out a project, not the least of which is access to able colleagues. Stigler has emphasized that access to colleagues is important because their conversation suggests fruitful topics, their criticism improves the quality of the research and their knowledge of the work increases the likelihood that it will have an impact on the larger community of scientists (Stigler 1986).

Pledge (1959) has claimed that the birthplace of scientists tends to have been clustered in certain locations. His suggestion, however, did not take into account general population density.

The distribution of Western polywater scientists (whether pro, con or neutral) by country is presented in Table A-19. The distribution of United States polywater scientists by state is presented in Table A-20.

The number of scientists for whom polywater had been suggested as a promising topic is reported in Table A-21. The importance of location in a research group in the choice of research project is highlighted by the following comments by scientists on who had influenced them in their choice of polywater as a project. The following comments are all those that were made on this issue.

Pro Sample

 (2) Senior scientists with my organization. I was a junior scientist at the time.

 (29) Ordered by W. A. Zisman. Zisman was Chemistry Department Superintendent at Naval Research Laboratory. I was two levels below him as a research chemist.

(185) Sponsor-Federal Agency.

(384) Co-worker in research at Xerox suggested concept of polywater vulnerable.

(241) Senior doing senior research project suggested we look into it. I knew how to use the equipment.

(41) Dr. James Kassner, Jr., was my faculty advisor.

(98) I first learned of it in the ONR newsletter and thought it looked interesting.

(107) I accepted their results (the Russians) as a challenge on a theoretical basis. Just as there are variations on the structure of ice, certainly I thought there could be variations in the structure of water. Maybe there are different structures. Certainly the structure of water is different inside of cells. But what causes this change? Perhaps it is the network of protein and polysaccharide chains inside the cell. Nevertheless, there *is* a difference in properties inside a cell as compared to outside that cell.

(202) Research advisor.

Neutral Sample

(122) Deryagin's publications suggested the topic was interesting. I have known him and respected his outstanding work for years.

Con Sample

(322) I was his postdoctoral fellow.

(386) Scientific Colleague interested in aqueous solutions at high pressure.

(154) I was a post doctoral associate at the time.

(172) I was a postdoc working for people who had money from the Office of Navel Research to attempt to prepare polywater and measure its properties if possible. Drs. Kassner and Zung obtained the grants.

(200) Supervisor

14. Family of Origin

A long tradition (e.g., Cattell and Brimhall 1921) suggests that scientific achievement may be related to the characteristics of the family of origin of the scientist. Some versions of this approach have been criticized by Stigler (1982b) and Coase (1984) using Mill and Marshall respectively as examples.

Becker (1981) has emphasized the trade-off between quantity and quality of children. I would expect, ceteris paribus, that children with few siblings would have a greater parental investment in their human capital, including (but not limited to) the kinds of human capital that would be useful to a productive scientist. Following up on a long-noted empirical regularity (e.g., Galton 1875), Zajonc and Marcus (1975) have argued that the allocation of time in the household will result in greater parental investment in the first born and, to a lesser extent, in the last born.

On the basis of these considerations, I would expect that scientific ability would be inversely related to the number of the scientist's siblings and would also be inversely related to the scientist having been a 'middle' child.

In Table A-23 I present data on the number of children (including the scientist himself) in the scientist's family of origin. Pro scientists seem to have come from larger families than con scientists. In Table A-24 I present data on the birth order of the scientists. No important differences in birth order appear between the various subsamples.

15. NUMBER OF CITATIONS

If one scientist cites the work of another scientist, then the first scientist is usually saying that the work of the first scientist (no matter how old) is still relevant to the current research. Thus, the number of citations that a scientist receives from her fellow scientists in a given year can be interpreted as a proxy for her stock of knowledge at the frontiers of the discipline (Diamond 1984). One interesting question is whether those with a larger stock of knowledge at the frontier will be more or less likely to choose polywater as a research project. Choosing a successful research project is partly due to good judgement and partly due to luck. If high citations are mainly due to good judgement, then we might expect that those with higher citations would be *less* likely to choose an ultimately-unsuccessful project such as polywater. On the other hand, if high citations are mainly due to lower levels of risk aversion, then we might expect that those with higher citations would be *more* likely to choose an ultimately-unsuccessful project such as polywater.

Looking at Table A-8, we find that citations in 1968 (AUTH68) is positively correlated: with having written on polywater (*either* pro *or* con), with having written a pro article on polywater, and with having written a highly visible pro article. This lends some credence to the view that those with higher citations have lower levels of risk aversion.

16. CONCLUSIONS

We presented a maximization-under-constraints framework that suggests that differences among scientists in their choice of research projects will depend on differences in tastes (most notably in risk aversion), and in differences in constraints (most notably the scientist's ability, time, information and research resources). Many of the hypotheses that have been commonly suggested to account for the behavior of scientists can be interpreted within this framework.

We examine several of these hypotheses using data from a particular episode in the recent history of science.

Several of the hypotheses have been discussed as though various characteristics of the scientist were exogenous. Some of these characteristics, at least partially, may be endogenous. Religious denomination, for instance, might in part be chosen for its compatibility with a life in science, rather than the other way around. The number of children that a scientist decides to have may be influenced by the intellectual (and therefore career) risks that she intends to take. Dealing with the potential partial endogeneity of many of these characteristics is an important objective of future work.

Some arbitrariness has been involved in the choice of hypotheses to consider. In retrospect others might equally well have been discussed. One would be the extent to which scientists consider the potential for practical applicability when choosing a research project. Polywater was perceived by many as having practical consequences in explaining the unusual behavior of water in a wide variety of contexts, including water in soil and in biological organisms.

Another hypothesis is that scientists sometimes choose a research project because they view it as being "fun." Several respondents mentioned this as a partial motive for the polywater research. Munevar (1981), pursuing a suggestion of Lorenz's, has suggested that much important scientific research is pursued as a form of play (1981, 66-69).

University of Nebraska at Omaha

TABLE A-1

Classification	Number of Authors Falling Under Classification
Excluded because non-Western	61
Excluded for having written only letters, review essays, articles for the popular press, or articles that only mention polywater in a trivial aside	115
Excluded for having written only articles written in some language besides English (number only includes those not already excluded because non-Western)	9
Excluded because I was unable to obtain any publications by them (most of those publications not obtained appear to be in a foreign language, so most of the scientists excluded in this category would have been excluded for falling in the previous category even if all publications had been obtained	52
Scientists who wrote only articles that were neutral	5
Scientists who wrote at least one article that was con and none that was pro	77
Scientists who wrote at least one article that was pro and none that was con	85
Scientists who wrote at least one article that was pro and at least one article that was con	11
Total Number of Authors in Gingold (1973) (sum of the above)	415

Classification of Authors of Entries on Gingold's List

TABLE A-2

Field of Employment	Number of Scientists	
Chemistry (academic)	87	(42.0)
Chemistry-Physics (academic)[b]	11	(5.3)
Chemistry-Specialized Applications[c] (academic)	7	(3.4)
Physics (academic)	12	(5.8)
Engineering (academic)	3	(1.4)
Biology and Biophysics (academic)	2	(1.0)
Other academic[d]	11	(5.3)
Government or private lab	58	(28.0)
Other	14	(6.8)
Missing	2	(1.0)
TOTAL	207	

[a]Indicates that both "physics" and "chemistry" were in the title of the unit of affiliation.

[b]Category includes: crystallography (3), chemical engineering (2), geochemistry (1), polymers (1), biochemistry and physics (1), and dental materials and surface chemistry (1).

[c]Category includes both: (a) specific academic affiliations other than those listed above and (b) academic affiliations too vague to classify. Note that under (b) are included some scientists who would with fuller information, be classed in one of the first four categories.

[d]Most of the information was obtained from the affiliation as listed on the article on polywater. Percentages for the nonmissing observations are given in parentheses beside the frequencies.

Field of Employment for Polywater Scientists at Time of Polywater Article[d]

TABLE A-3

Variable	N	Mean	Standard Deviation	Minimum	Maximum
Age in 1971	41	40.1	10.3	23	79
Year of BS	33	1954.5	9.4	1923	1970
Year of Ph.D.	46	1959.3	9.5	1926	1979
Citations in 1968	97	35.5	131.1	0	1000
Citations in 1969	96	35.3	131.9	0	1087
Citations in 1974	97	44.1	165.4	0	1546
Citations in 1981	96	40.6	118.1	0	1016
Articles in 1970	97	3.1	3.9	0	20
Articles in 1971	97	3.2	3.6	0	20
Articles in 1972	97	2.3	3.2	0	17
Articles in 1973	97	2.0	2.8	0	14
Articles in 1974	97	2.0	3.0	0	10

Description Statistics for Pro Subsample (N=98)

TABLE A-4

Variable	N	Mean	Standard Deviation	Minimum	Maximum
Age in 1971	20	40.9	10.0	22	60
Year of BS	19	1951.9	10.6	1931	1970
Year of Ph.D.	26	1960.5	10.3	1935	1978
Citations in 1968	38	20.3	31.1	0	100
Citations in 1969	38	26.9	41.6	0	151
Citations in 1974	38	34.1	46.9	0	176
Citations in 1981	38	43.2	56.6	0	245
Articles in 1970	38	4.4	4.9	0	26
Articles in 1971	38	3.2	3.6	0	14
Articles in 1972	38	2.5	3.3	0	14
Articles in 1973	38	3.7	4.4	0	15
Articles in 1974	38	2.8	3.5	0	15

Descriptive Statistics for Science Subsample (N=38)

TABLE A-5

Variable	N	Mean	Standard Deviation	Minimum	Maximum
Age in 1971	24	40.3	9.3	26	64
Year of BS	23	1953.1	9.5	1928	1967
Year of Ph.D.	33	1962.1	8.1	1934	1977
Citations in 1968	76	15.2	29.0	0	128
Citations in 1969	76	16.7	31.2	0	135
Citations in 1974	76	18.1	34.1	0	198
Citations in 1981	76	25.5	47.1	0	277
Articles in 1970	76	2.4	3.5	0	20
Articles in 1971	76	2.5	3.5	0	21
Articles in 1972	76	2.1	3.8	0	21
Articles in 1973	76	1.7	3.3	0	20
Articles in 1974	76	2.2	3.9	0	28

Descriptive Statistics for Con Subsample (N = 76)

TABLE A-6

Number of Publications on Polywater per Scientist	Number of Scientists in Sample Who Wrote Number of Publications in First Column	Names of Scientists in Sample Who Wrote Number of Publications in First Column
1	102	
2	33	
3	13	
4	12	
5	6	Davis, R. E.; Drost-Hansen, W.; Jakobsen, R. J.; Lingertat, H.; Rao, C. N. R.; Rousseau, D. L.
6	4	Bradspies, J. I.; Everett, D. H.; Finney, J. C.; Howell, B. F.
7	3	Entine, G.; Horne, R. A.; Killman, P. A.
8	2	Brummer, S. B.; Lippincott, E. R.
9	1	Adlfinger, K. H.
10	1	Allen, L. C.
11	2	Abendroth, R. P.; Peschel, G.
	179	

[a]"Publications" here may include letters to the editor, review articles and articles for the popular press. To be included in the dample, however, a scientist had to have written at least one article on polywater.

**Number of Publications[a] on Polywater
Written by Polywater Scientists**

TABLE A-7

POLY	1 = scientist wrote on polywater; 0 = scientist did not write on polywater
SPEED	year of Ph.D. minus year of BS
RANK69 (RANK75)	0 = lecturer in 1969 (1975); 1 = instructor in 1969 (1975); 2 = assistant professor in 1969 (1975); 3 = associate professor in 1969 (1975); 4 = full professor in 1969 (1975)
NUMARTPW	number of publications on polywater
PRO	1 = pro on polywater 0 = con, neutral or did not write on polywater
PROCLUST	1 = highly visible pro-polywater 0 = everyone else
AUTH68 (AUTH74)	citations received by scientist in 1968 (1974)
POST69AR	number of non-polywater articles written from 1970 through 1974
AGEIN71	1971-(year of birth)

Definition of Variables Used in Correlation Table

TABLE A-8

	POLY	SPEED	RANK69	RANK75	NUMARTPW	AGEIN71
POLY	1.000	0.120	0.086	0.048	0.415	-0.023
	0.00	0.29	0.54	0.74	0.00	0.84
	216	78	53	49	216	87
SPEED	0.120	1.000	-0.403	0.062	0.136	0.197
	0.29	0.00	0.01	0.69	0.24	0.08
	78	78	46	45	78	78
RANK69	0.086	-0.403	1.000	0.905	0.086	0.253
	0.54	0.01	0.00	0.00	0.54	0.08
	53	46	53	43	53	50
RANK75	0.048	0.062	0.905	1.000	0.022	0.293
	0.74	0.69	0.00	0.00	0.88	0.04
	49	45	43	49	49	49
NUMARTPW	0.415	0.136	0.086	0.022	1.000	0.021
	0.00	0.24	0.54	0.88	0.00	0.85
	216	78	53	49	216	87
AGEIN71	-0.023	0.197	0.253	0.293	0.021	1.000
	0.84	0.08	0.08	0.04	0.85	0.00
	87	78	50	49	87	87
PRO	0.417	0.017	0.209	0.124	0.378	-0.043
	0.00	0.88	0.13	0.39	0.00	0.69
	216	78	53	49	216	87
PROCLUST	0.156	-0.016	0.159	-0.146	0.210	-0.081
	0.02	0.89	0.25	0.32	0.00	0.46
	216	78	53	49	216	87
AUTH68	0.025	-0.094	0.183	0.139	0.035	0.143
	0.71	0.41	0.19	0.34	0.61	0.19
	216	78	53	49	216	87
AUTH74	-0.006	-0.090	0.163	0.098	0.019	0.092
	0.93	0.43	0.24	0.50	0.79	0.40
	216	78	53	49	216	87
POST69AR	-0.166	-0.194	0.124	0.041	0.015	0.016
	0.01	0.09	0.38	0.78	0.82	0.88
	216	78	53	49	216	87
PROFAGE	0.014	-0.201	0.480	0.389	0.078	0.878
	0.88	0.08	0.00	0.01	0.43	0.00
	107	78	52	49	107	83

[a]Correlation Coefficients/Prob > | R | Under HO:RHO=0/Number of Observations.

Correlation Coeffieients for Selected Variables[a]

TABLE A-8 (continued)

	PRO	PROCLUST	AUTH68	AUTH74	POST69AR	PROFAGE
POLY	0.417	0.156	0.025	-0.006	-0.166	0.014
	0.00	0.02	0.71	0.93	0.01	0.88
	216	216	216	216	216	107
SPEED	0.017	-0.016	-0.094	-0.090	-0.194	-0.201
	0.88	0.89	0.41	0.43	0.09	0.08
	78	78	78	78	78	78
RANK69	0.209	0.159	0.183	0.163	0.124	0.480
	0.13	0.25	0.19	0.24	0.38	0.00
	53	53	53	53	53	52
RANK75	0.124	-0.146	0.139	0.098	0.041	0.389
	.039	0.32	0.34	0.50	0.78	0.01
	49	49	49	49	49	49
NUMARTPW	0.378	0.210	0.035	0.019	0.015	0.078
	0.00	0.00	0.61	0.79	0.82	0.43
	216	216	216	216	215	107
AGEIN71	-0.043	-0.081	0.143	0.092	0.016	0.878
	0.69	0.46	0.19	0.40	0.88	0.00
	87	87	87	87	87	83
PRO	1.000	0.281	0.102	0.090	-0.031	0.089
	0.00	0.00	0.13	0.19	0.65	0.36
	216	216	216	216	216	107
PROCLUST	0.281	1.000	0.099	0.035	-0.060	0.018
	0.00	0.00	0.15	0.60	0.38	0.85
	216	216	216	216	216	107
AUTH68	0.102	0.099	1.000	0.909	0.341	0.224
	0.13	0.15	0.00	0.00	0.00	0.02
	216	216	216	216	216	107
AUTH74	0.090	0.035	0.909	1.000	0.427	0.161
	0.19	0.60	0.00	0.00	0.00	0.10
	216	216	216	216	216	107
POST69AR	-0.031	-0.060	0.341	0.427	1.000	0.136
	0.65	0.38	0.00	0.00	0.00	0.16
	216	216	216	216	216	107
PROFAGE	0.089	0.018	0.224	0.161	0.136	1.000
	0.36	0.85	0.02	0.10	0.16	0.00
	107	107	107	107	107	107

Correlation Coefficients for Selected Variables (continued)

TABLE A-9

Journal, Year and First page of Article	Total Citations in 1970 to Cocited Article (if 16 or more)	Total Citations in 1971 to Cocited Article (if 16 or more)	Scientists Who Coauthored Article	Pages of Frank's Judgment that Article Was Pro
Science, 1970, p. 1443	—	27	Allen, L.C. Kollman, P.A.	p. 90
Chem. Ind. London, 1969, p. 686	24	17	Bellamy, L.J. Osborn, A.R. Lippincott, E.R. Brady, A.R.	pp. 68-69
Nature, 1969, p. 865	17	—	Bolander, R.W. Kassner, I.L. Zung, J.T.	p. 65
Science, 1970, p. 865	—	16	Castellion, G.A. Grabar, D.G. Hession, J. Burkhard, H.	—
Science, 1969, p. 1482	58	46	Lippincott, E.R. Stromberg, R.R. Grant, W.H. Cessac, G.L.	p. 70
Science, 1970, p. 51	—	32	Page, T.F. Jakobsen, R.J. Lippincott, E.R.	—
Science, 1970, p. 171	—	17	Petsko, G.A.	—
Nature, 1969, p. 159	24	21	Willis, E. Rennie, G.K. Smart, C. Pethica, B.A.	pp. 61-62

[a]An article was judged to be highly visible if it was a member of a group of articles that were frequently cited together. Articles not mentioned in the Frank's book (1981), were examined directly to judge whether they were pro, con, or neutral.

Data on Highly Visible Pro Articles on Polywater[a]

TABLE A-10

Name	Age in 1971	Year of Ph.D.	Citations in 1968	Citations in 1974	Citations in 1981	Location in 1969-71	Location in 1979-83
Allen, L.C.	45	1957	49	66	37	Princeton	Princeton
Bandy, A.R.	31	1968	0	5	2	Old Dom. Univ.	Drexel
Bellamy, L.J.	—	—	809	488	469	—	1981-Univ. of Birmingham (England)
Bolander, R.	31	1969	0	4	2	General Motors Institute	Same
Burkhard, H.	—	—	0	1	3	—	1982-Deutsch Bundespost FT2 (Fed. Rep. Ger.)
Castellion, G.A.	—	1956	2	0	2	—	—
Cessac, G.L.	—	—	0	2	3	—	—
Grabaer, D.G.	—	—	1	2	0	—	—
Grant, W.H.	38	1968	3	3	12	Nat. Bureau Standards	Same
Hessian, J.	—	—	0	0	0	—	—
Jakobsen, R.J.	42	—	37	27	79	Battelle Cols. Labs	Same
Kassner, I.L.	40	1957	6	5	15	Univ. of Missouri-Rolla	Same
Kollman, P.A.	27	1970	0	160	116	Univ. of Cal. SF	Same
Lippincott, E.R.	51	1947	120	112	74	Univ. Maryland	Same
Osborn, A.R.	—	—	6	4	2	—	—
Page, T.F.	—	—	14	21	45	Univ. of IL-Urbana	Same
Pethica, B.A.	45	1949	44	36	32	—	1981-Clarkson Col. Technology, NY
Petsko, G.A.	23	1973	0	2	21	Wayne State Univ.	MIT(81)
Rennie, G.K.	—	—	1	1	3	—	—
Smart, C.	—	—	17	5	17	—	1983-under Smart, C.C. Univ. of Edinburgh (Scotland)
Stromberg, R.R.	—	1951	24	40	18	—	—
Willis, E.	—	—	3	2	3	—	1981-City Hospital Copenhagen (Denmark)
Zung, J.T.	41	1960	8	2	2	Univ. Missouri-Rolla	Same

Age, Citation and Location Data on Pro Polywater Scientists

TABLE A-11

Journal, Year and First Page of Article	Total Citations in 1970 to Co-cited Article (if 16 or more)	Total Citations in 1971 to Co-cited Article (if 16 or more)	Scientists Who Coauthored Article	Pages of Frank's Judgment That Article Was Con
Nature, 1970, p. 1033	—	19	Everett, D.H. Haynes, J.M. McElroy, P.J.	pp. 85-86
Science, 1970, p. 48	—	22	Rabideau, S.W. Florin, A.E.	p. 130
Science, 1970, p. 1715	—	38	Rousseau, D.L. Porto, S.P.S.	pp. 101, 135

[a]An article was judged to be highly visible if it was a member of a group of articles that were frequently cited together.

Data on Highly Visible Con Articles on Polywater[a]

TABLE A-12

Name	Age in 1971	Year of Ph.D.	Citations in 1968	Citations in 1974	Citations in 1981	Location in 1969-71	Location in 1979-83
Everett, D.H.	—	—	148	119	155	—	—
Florin, A.E.	—	—	3	1	3	—	—
Haynes, J.M.	—	—	7	3	9	—	—
McElroy, P.J.	—	—	0	0	3	—	—
Porto, S.P.S.	—	—	51	37	26	—	—
Rabideau, S.W.	—	—	36	18	15	—	—
Rousseau, D.L.	31	1967	5	15	0	Univ. of Southern Cal.	Bell Labs (1981)

Age, Citation and Location Data on Con Polywater Scientists

TABLE A-13

Subsample	Number in Subsample	Number Responded in Subsample	Percentage of Subsample Who Responded
Con	77	11	14.3
Neutral	5	1	20.0
Pro	96	12	12.5
Wrote on Polywater	178	24	13.5
Highly Visible Pro	23	4	17.4
Science	38	9	23.7

Questionnaire Response Rates for Subsamples

TABLE A-14

Responses	Con	Neutral	Pro	Wrote on Polywater	Highly Visible
A Very Risky Topic	3 (30.0)	—	3 (25.0)	6 (26.1)	2 (50.0)
A Topic of Greater Than Average Riskiness	2 (20.0)	—	2 (16.7)	4 (17.4)	1 (25.0)
A Topic About Average in Riskiness	1 (10.0)	—	3 (25.0)	4 (17.4)	—
A Safe Topic	3 (30.0)	—	1 (8.3)	4 (17.4)	—
Did Not Respond	1 (100.0)	1 (10.0)	3 (25.0)	5 (21.7)	1 (25.0)

Consideration of Polywater as a Risky Topic

TABLE A-15

Responses	Con	Neutral	Pro	Wrote on Polywater	Highly Visible	Science
Zero	—	—	1 (7.1)	1 (4.5)	—	—
One	—	—	1 (7.1)	1 (4.5)	—	—
Two	1 (14.3)	—	4 (28.6)	5 (22.7)	—	3 (60.0)
Three	6 (85.7)	—	3 (21.4)	9 (40.9)	—	2 (40.0)
Four	—	1 (100.0)	3 (21.4)	4 (18.2)	2 (66.7)	—
Five	—	—	1 (7.1)	1 (4.5)	1 (33.3)	—
More Than Five	—	—	1 (7.1)	1 (4.5)	—	—

Number of Children

TABLE A-16

Responses	Con	Neutral	Pro	Wrote on Polywater	Highly Visible Pro	Science
less than $10,000	2 (18.2)	—	2 (16.7)	4 (16.7)	1 (25.0)	1 (11.1)
$10,000 - $15,000	1 (9.1)	—	1 (8.3)	2 (8.3)	1 (25.0)	1 (11.1)
$15,000 - $20,000	—	—	4 (33.3)	4 (16.7)	1 (25.0)	1 (11.1)
$20,000 - $25,000	2 (18.2)	—	3 (25.0)	5 (20.8)	1 (25.0)	1 (11.1)
$25,000 - $30,000	1 (9.1)	1 (100.0)	1 (8.3)	3 (12.5)	—	1 (11.1)
$30,000 - $35,000	1 (9.1)	—	1 (8.9)	2 (8.3)	—	—
Did Not Respond	4 (36.44)	—	—	4 (16.7)	—	4 (44.4)

Monetary Income in 1971

TABLE A-17

Subsample	Lecturer-Instructor	Assistant Professor	Associate Professor	Full Professor	Missing
Con	2 (18.2)	2 (18.2)	2 (18.2)	5 (46.5)	65
Neutral	0	0	1 (50.0)	1 (50.0)	3
Pro	0	3 (12.5)	8 (33.3)	13 (54.2)	74
Wrote on Polywater	2 (5.4)	5 (13.5)	11 (29.7)	19 (51.4)	142
Highly Visible Pro-Polywater	0	0	2 (33.3)	4 (66.7)	17
Science Non-Polywater	2 (12.5)	3 (18.7)	2 (12.5)	9 (56.2)	22

[a]Percentages for the nonmissing observations are given in parentheses underneath the frequencies.

Academic Rank in 1969[a]

TABLE A-18

Responses	Con	Neutral	Pro	Wrote on Polywater	Highly Visible Pro	Science
Agnostic or Atheist	2 (18.2)	—	3 (25.0)	5 (20.8)	1 (25.0)	2 (22.2)
Catholic	1 (9.1)	—	1 (8.3)	2 (8.3)	—	1 (11.1)
Jewish	1 (9.1)	—	—	1 (4.2)	—	—
Protestant	4 (36.4)	1 (100.0)	5 (41.7)	10 (41.7)	2 (50.0)	2 (22.2)
Other	1 (9.1)	—	3 (25.0)	4 (16.7)	1 (25.0)	2 (22.2)
Did Not Respond	2 (18.2)	—	—	2 (8.3)	—	2 (22.2)

Current Religious Affiliation

TABLE A-19

Country	Number of Scientists
Australia	9
Belguim	2
Canada	10
England	19
France	1
India	1
Israel	1
Italy	6
Scotland	2
USA	101
West Germany	2
TOTAL	159

**Geographical Distribution by Country of
Polywater Scientists at Time of
Publication of Polywater Articles**

TABLE A-20

State	Number of Scientists
California	11
Connecticut	4
Florida	1
Illinois	2
Indiana	3
Louisiana	5
Massachusetts	8
Maryland	4
Michigan	1
Missouri	11
New Hampshire	3
New Jersey	7
New Mexico	2
New York	7
North Carolina	2
North Dakota	3
Ohio	8
Pennsylvania	9
Tennessee	2
Texas	2
Washington, D.C.	5
Wisconsin	1
TOTAL	101

Geographical Distribution of U.S.
Polywater Scientists at Time of
Publication of Polywater Articles

TABLE A-21

Responses	Con	Neutral	Pro	Wrote on Polywater	Highly Visible Pro
Yes	5 (45.5)	1 (100.0)	3 (25.0)	9 (37.5)	2 (50.0)
No	5 (45.5)	—	9 (75.0)	14 (58.3)	2 (50.0)
I Don't Remember	1 (9.1)	—	—	1 (4.2)	—

Polywater Suggested as a Promising Topic of Research

TABLE A-22

Responses	Con	Neutral	Pro	Wrote On Polywater	Highly Visible Pro	*Science*
One	1 (11.1)	—	2 (12.5)	3 (11.5)	—	1 (14.3)
Two	5 (55.6)	1 (100.0)	3 (18.8)	9 (34.6)	1 (25.0)	3 (42.9)
Three	2 (22.2)	—	6 (37.5)	8 (30.8)	1 (25.0)	—
Four	1 (11.1)	—	3 (18.8)	4 (15.4)	1 (25.0)	1 (14.3)
Five	—	—	1 (6.3)	1 (3.8)	—	2 (28.6)
. . Eight	—	—	1 (6.3)	1 (3.8)	1 (25.0)	—

Number of Children in Family of Origin

TABLE A-23

Responses	Con	Neutral	Pro	Wrote on Polywater	Highly Visible Pro	Science
First-born	7 (77.8)	1 (100.0)	8 (50.0)	16 (61.5)	1 (25.0)	5 (71.4)
Second-born	1 (11.1)	—	5 (31.3)	6 (23.1)	1 (25.0)	1 (14.3)
Third-born	1 (11.1)	—	—	1 (3.8)	1 (25.0)	—
Fourth-born	—	—	1 (6.3)	1 (3.8)	—	1 (14.3)
Fifth-born	—	—	1 (6.3)	1 (3.8)	—	—
Sixth-born or Later	—	—	1 (6.3)	1 (3.8)	1 (25.0)	—

Birth Order of Scientist

Notes

*The research reported in this paper was partially supported by a grant from the National Science Foundation and a grant-in-aid from the College of Social and Behavioral Sciences of the Ohio State University. I am especially grateful for the detailed comments from Luis Locay, Hajime Miyazaki, Brian A. Pethica, Aloysius Siow and Larry Stern. I have also received useful comments from Felix Franks and Harmon Maher. Richard Klimoski provided valuable suggestions on polling techniques. I am grateful to Gregory Armotrading, Mark Chapinski, Jack Julian, Lisa Knazek, Bret Mizer, Maureen Ogle, Christopher Smith, James Thomas, William Tisch, Ann Wertz and Kathryn Williams for able research assistance. Henry Small of the Institute for Scientific Information graciously provided me with 1970 and 1971 cocitation cluster data. Irving Klotz first suggested to me the appropriateness of the polywater episode as an example of a mistake. An earlier version of the paper was presented at the 1986 annual meeting of the American Economic Association in New Orleans. A couple of the paragraphs on the background of the polywater episode first appeared in my paper "The Polywater Episode and the Appraisal of Theories."

1. The layperson's externalist view of the scientist has never been more enthusiastically expressed than by Adam Smith in the *Theory of Moral Sentiments*:

> Mathematicians and natural philosophers, from their independency upon the public opinion, have little temptation to form themselves into factions and cabals, either for the support of their own reputation, or for the depression of that of their rivals. They are almost always men of the most amiable simplicity of manners, who live in good harmony with one another, are the friends of one another's reputation, enter into no intrigue in order to secure the public applause, but are pleased when their works are approved of, without being either much vexed or very angry when they are neglected. (Smith 1976, 125)

2. For a defense of the claim that the usual interpretation of Kuhn is closer to the text than is Kuhn and Merton's interpretation, see Diamond 1978, 5-7.

3. A useful general discussion of the internal-external debate can be found in Barnes 1974.

4. Gingold lists 438 total entries, but his entry 240 is identical to his entry 30. Hence he actually has only 437 total entries.

5. Several commentators on polywater have overestimated the number of scientists actively involved in the polywater episode. Bennion and Neuton, for instance, claim that "[a] conservative estimate would place the number at no less than 10^5 for those having the immediate potential of becoming involved in polywater research. Only 430 of these, or less than 0.5 percent, actually became infected, i.e., published results of their own investigations" (1976, 54). Grove, for another instance, writes that "...by 1972 some 400 scientists had published results about this exciting new substance" (Grove 1985, 133). The Bennion and Neuton claim is based on the Gingold list. The source of the Grove claim is unclear. A casual reading of the Gingold bibliography leads to an overestimate of the quantity of research on polywater because the bibliography includes letters, articles in the popular press, review essays, and articles that, although they mention polywater (albeit in a very casual way), are written on another topic.

References

American Chemical Society. (1977), *Directory of Graduate Research 1977*. Washington, DC: American Chemical Society.

Arrow, K. J. (1971), *Essays in the Theory of Risk-Bearing*. Chicago: Markham.

Barnes, B. (1974), *Scientific Knowledge and Sociological Theory*. London: Routledge & Kegan Paul.

Becker, G. S. (1981), *A Treatise on the Family*. Cambridge, MA: Harvard University Press.

Bennion, B. C. and Newton, L. A. (1976), "The Epidemology of Research on 'Anomalous Water,'" *Journal of the American Society for Information Science* 27: 53-56.

Blackmore, J. T. (1978), "Is Planck's 'Principle' True?," *British Journal for the Philosophy of Science* 29: 347-349.

Brenner, R. (1983), *History—The Human Gamble*. Chicago: University of Chicago Press.

Brush, S. G. (1976), *The Kind of Motion I Call Heat*, Book 1, in E. W. Montroll and J. L. Lebowitz (eds.), *Studies in Statistical Mechanics*, vol. 6. New York: North-Holland.

Cantor, G. N. (1975), "The Edinburgh Phrenology Debate: 1803-1823," *Annals of Science 32*: 195-218.

Carmichael, H. L. (1988), "Incentives in Academics: Why Is There Tenure?," *Journal of Political Economy* 96 (no. 3): 453-472.

Cattell, J. M. and Brimhall, D. R. (1921), "Families of American Men of Science," in *American Men of Science*. Garrison, NY: The Science Press, pp. 781-808.

Coase, R. H. (1984), "Alfred Marshall's Mother and Father," *History of Political Economy* 16: 519-527.

Cole, J. R. and Cole, S. (1973), *Social Stratification in Science*. Chicago: University of Chicago Press.

Crane, D. (1972), *Invisible Colleges*. Chicago: University of Chicago Press.

Deryagin, B. (1983), "Polywater Reviewed," *Nature* 301: 9-10.

Diamond, A. M., Jr. (1978), "Science as a Rational Enterprise," Ph.D. dissertation, (University of Chicago Department of Philosophy.)

———. (1980), "Age and the Acceptance of Cliometrics," *Journal of Economic History* 40: 838-841.

———. (1982), "Avery's 'Neurotic Reluctance,'" *Perspectives in Biology and Medicine* 26: 132-136.

———. (1984), "An Economic Model of the Life-Cycle Research Productivity of Scientists," *Scientometrics* 6 (no. 3): 189-196.

———. (1988a), "The Polywater Episode and the Appraisal of Theories," in A. Donovan, L. Laudan and R. Laudan (eds.), *Scrutinizing Science: Empirical Studies of Scientific Change*. Dordrecht: Kluwer Academic, pp. 181-98.

———. (1988b), "Science as a Rational Enterprise," *Theory and Decision* 24: 147-167.

Diamond, A. M., Jr. and Locay, L. (1989), "Investment in Sister's Children as Behavior Towards Risk," *Economic Inquiry* 27 (no. 4): 719-735.

_____. (1986), "What is a Citation Worth?," *The Journal of Human Resources* 21: 200-215.

Edge, D. (1977), "The Sociology of Innovation in Modern Astronomy," *Quarterly Journal of the Royal Astronomical Society* 18: 326-339.

Eisenberg, D. (1981), "A Scientific Gold Rush," *Science* 213 (September 4): 1104-1105.

Fleming, D. (1968), "Emigre Physicists and the Biological Revolution," *Perspectives in American History* 2: 152-189.

Franks, F. (1968), "Effects of Solutes and Surfaces on the Structural Properties of Liquid Water," in L. Bolis and B. A. Pethica (eds.), *Membrane Models and Formation of Biological Membranes*. Proceedings of the 1967 meeting of the International Conference on Biological Membranes. New York: Wiley & Sons, pp. 43-51.

_____. (1981), *Polywater*. Cambridge, MA: The M.I.T. Press.

Freeman, S. (1977), "Wage Trends as Performance Displays Productive Potential: A Model and Application to Academic Early Retirement," *Bell Journal of Economics* 8: 419-443.

Friedman, M. (1953), "The Methodology of Positive Economics," in *Essays in Positive Economics*. Chicago: University of Chicago Press, pp. 3-43.

Galton, F. (1875), *English Men of Science*. New York: Macmillan.

Garfield, E. (chair), (1980), *Science Citation Index 1979*. Philadelphia: Institute for Scientific Information.

_____. (1983), "Computer-Aided Historiography—How ISI Uses Cluster Tracking to Monitor the 'Vital Signs' of Science," in *Essays of an Information Scientist*, vol. 5, Philadelphia: I.S.I. Press, pp. 473-83.

Gieryn, T. F. and Hirsh, R. F. (sic) (1983), "Marginality and Innovation in Science," *Social Studies of Science* 13: 87-106.

Gingold, M. P. (1973), "Anomalous Water: General Review," *Societe Chimique de France, Bulletin*, Pt. 1: 1629-1644.

_____. (1974), "L'eau Anomale: Histoire d'un Artefact," *La Recherche* 5: 390-393.

Goffman, W. and Warren, K. S. (1980), *Scientific Information Systems and the Principle of Selectivity*. New York: Praeger.

Gould, S. J. (1981), "Ice Nine, Russian-Style," *The New York Times Book Review* 130 (no. 45,056): 7, 15.

Grove, J. W. (1985), "Rationality at Risk: Science against Pseudoscience," *Minerva* 23: 216-240.

Hagstrom, W. O. (1965), *The Scientific Community*. New York: Basic Books.

Hardy, K. R. (1974), "Social Origins of American Scientists and Scholars," *Science* 185: 497-506.

Hargens, L. L. (1988), "Scholarly Consensus and Journal Rejection Rates," *American Sociological Review* 53: 139-151.

Harris, M. and Weiss, Y. (1984), "Job Matching with Finite Horizon and Risk Aversion," *Journal of Political Economy* 92: 758-779.

Hasted, J. B. (1971), "Water and 'Polywater,'" *Contemporary Physics* 12: 133-152.

Howell, B. F. (1971), "Anomalous Water: Fact or Figment," *Journal of Chemical Education* 48: 663-667.

Hull, D. L. (1978), "Altruism in Science: A Sociobiological Model of Co-operative Behaviour Among Scientists," *Animal Behavior 26*: 685-697.

Hull, D. L., Tessner, P. D. and Diamond, A. M. (1978), "Planck's Principle," *Science* 202: 717-723.

Ito, T. and Kahn, C. (1986), "Why is There Tenure?" Discussion Paper No. 228, Center for Economic Research, University of Minnesota.

Jacques Cattell Press (ed.), (1979), *American Men and Women of Science*, 14th ed. New York: Bowker.

Kahn, C. and Huberman, G. (1988), "Two-sided Uncertainty and 'Up-or-Out' Contracts," *Journal of Labor Economics* 6: 423-444.

Klotz, I. M. (1980), "The N-Ray Affair," *Scientific American 242*: 168-175.

Kolata, G. (1984), "Surprise Proof of an Old Conjecture," *Science* 225: 1006-1007.

_____ . (1986), "What Does It Mean to Be Random?" *Science* 231: 1068-1070.

Kollman, P. A. and Allen, L. C. (1972), "The Theory of the Hydrogen Bond," *Chemical Reviews* 72: 283-303.

Kuhn, T. S. (1970), *The Structure of Scientific Revolutions*, 2d ed. Chicago: University of Chicago Press.

Langmuir, I. (1989), "Pathological Science," *Physics Today 42*: 36-48.

Laudan, L. (1977), *Progress and Its Problems*. Berkeley and Los Angeles: University of California Press.

Long, J. S. (1978), "Productivity and Academic Position in the Scientific Career," *American Sociological Review* 43: 889-908.

MacRoberts, M. H. and MacRoberts, B. R. (1984), "The Negational Reference: or the Art of Dissembling," *Social Studies of Science* 14: 91-94.

McCarty, M. (1980), "Reminiscences of the Early Days of Transformation," *Annual Review of Genetics* 14: 1-15.

McClelland, D. (1976), *The Achieving Society*. New York: Irvington Press.

McCloskey, D. N. (1985), *The Rhetoric of Economics*. Madison: University of Wisconsin Press.

Merton, R. K. ([1936] 1970), *Science, Technology and Society in Seventeenth-Century England*. New York: Fertig.

Metzger, N. (1970), "Polywater Boils," *Chemical and Engineering News* (November 9): 9.

Morin, R. A. and Suarez, A. F. (1983), "Risk Aversion Revisited," *The Journal of Finance* 38: 1201-1216.

Mulkay, M. J., Gilbert, G. N. and Woolgar, S. (1975), "Problem Areas and Research Networks in Science," *Sociology* 9: 187-203.

Mullins, N. C.; Hargens, L. L.; Hecht, P. K.; and Kick, E. L. (1977), "The Group Structure of Cocitation Clusters: A Comparative Study," *American Sociological Review* 42: 552-556.

Munevar, G. (1981), *Radical Knowledge*. Indianapolis: Hackett.

Pethica, B. A. (1982), "Review of *Polywater*," *Journal of Colloid and Interface Science* 88: 607.

Planck, M. (1949), *Scientific Autobiography and Other Papers*. New York: Philosophical Library, pp. 33-34.

Pledge, H. T. ([1939] 1959), *Science Since 1500*. New York: Harper & Brothers.

Pollock, G. L. (Project Editor) (1984), *Chemical Research Faculties: An International Directory*. Washington, DC: American Chemistry Society.

Rubin, P. H. and Paul II, C. W. (1979), "An Evolutionary Model of the Taste for Risk," *Economic Inquiry* 17: 585-595.

Sabin, J. R.; Harris, R. E.; Archibald, T. W.; Kollman, P. A.; Allen, L. C. (1970), *Ab Initio* MO-SCF Calculation on a Model of Anomalous Water, *Theoretica Chimica Acta* (Berl.) 18: 235-238.

Siow, A. (1979), "A Model of Academic Tenure." Preliminary draft presented at Applications of Economics Workshop, The University of Chicago.

Small, H. G. and Griffith, B. C. (1974), "The Structure of Scientific Literatures, I: Identifying and Graphing Specialties," *Science Studies* 4: 17-40.

Smith, A. (1976), *The Theory of Moral Sentiments*. London: Oxford University Press.

Stigler, G. J. (1982a), "The Literature of Economics: The Case of the Kinked Oligopoly Demand Curve," in *The Economist as Preacher and Other Essays*. Chicago: University of Chicago Press, pp. 223-43.

————. (1982b), "The Scientific Uses of Scientific Biography, with Special Reference to J. S. Mill," in *The Economist as Preacher and Other Essays*. Chicago: University of Chicago Press, pp. 86-97.

————. (1983), "Nobel Lecture: The Process and Progress of Economics," *Journal of Political Economy* 91: 529-545.

————. (1986), "George J. Stigler," in W. Breit and R. W. Spencer, (eds.), *Lives of the Laureates: Seven Nobel Economists*. Cambridge, MA: MIT Press, pp. 93-111.

Studer, K. E. and Chubin, D. E. (1980), *The Cancer Mission: Social Contexts of Biomedical Research*. Sage Library of Social Research 103. Beverly Hills: Sage.

Telser, L. G. (1974), "Advertising and the Consumer," in Y. Brozen (ed.), *Advertising and Society*. New York: New York University Press, 25-42.

Toulmin, S. (1972), *Human Understanding*. Princeton: Princeton University Press.

Wallis, R. (ed.), (1979), *On the Margins of Science: The Social Construction of Rejected Knowledge*. Sociological Review Monograph 27. Keele, Staffordshire: University of Keele.

Watson, J. D. (1968), *The Double Helix*. New York: The New American Library.

Weber, M. ([1930] 1958), *The Protestant Ethic and the Spirit of Capitalism*. New York: Charles Scribner's Sons.

Weiss, Y. and Lillard, L. (1982), "Output Variability, Academic Labor Contracts and Waiting Time for Promotion," in R. G. Ehrenberg (ed.), *Research in Labor Economics 5*: 157-188.

Zajonc, R. B. and Markus, G. B. (1975), "Birth Order and Intellectual Development," *Psychological Review* 82: 74-88.

Zuckerman, H. (1977), "Deviant Behavior and Social Control in Science," in Edward Sagarin (ed.), *Deviance and Social Change*. London: Sage, pp. 87-138.

_____ . (1979), "Theory Choice and Problem Choice in Science," in J. Gaston (ed.), *Sociological Perspectives on Science*. San Francisco: Jossey-Bass, pp. 65-95.

Contributors

Pnina G. Abir-Am
History of Science
Johns Hopkins University

James Allen
Department of Philosophy
University of Pittsburgh

Frank Arntzenius
Department of Philosophy
University of Southern California

Arthur M. Diamond, Jr.
Department of Philosophy
University of Nebraska at Omaha

Edward J. Green
Department of Economics
University of Minnesota

Tamara Horowitz
Department of Philosophy
University of Pittsburgh

Allen I. Janis
Department of Physics and Astronomy
University of Pittsburgh

Peter King
Department of Philosophy
Ohio State University

James V. Maher
Department of Physics and Astronomy
University of Pittsburgh

Keith A. Moss
Chicago, Illinois